T0358638

The Lean Lifestyle Strategy for Businesses

In a world of increasing variability and complexity, companies seem to persist in using outdated and inadequate organizational models and work patterns. Despite the available technologies and the most innovative time management techniques, we work more and more, with less results and more stress.

Lean Lifestyle Strategy addresses the key problem of every professional, manager, and entrepreneur leading companies of all sizes: how to combine the need to produce more and more results, in less and less time, working better, and leading, at the same time, a lifestyle that generates prosperity and well-being.

The time is ripe for lean work to become a strategy to achieve a true work–life balance and express the best of people in the company without having to choose between operational efficiency or personal fulfilment.

In this book, in addition to numerous examples, tools, and step-by-step methodologies, useful to begin to apply independently the principles of the Lean Lifestyle Strategy, you will find collected the testimonies of entrepreneurs and managers who reveal the "behind-the-scenes" of successful cases in this new direction, including Campari, Cromology, ELT Group, Ferretti Group, Labomar, Lucchini RS, Marcegaglia, Orogel, Poste Italiane, Sammontana, Siemens Italia, Stanley Black & Decker, and Streparava.

Through the evolution of the way we work and do business, it is possible to leave a tangible mark that starting from the company boundaries touches ourselves and the people who work with us, until positively influencing the society and the environment where we live.

The Lean Lifestyle Strategy for Businesses

An Operational Method for Entrepreneurs and Managers

Luciano Attolico

A PRODUCTIVITY PRESS BOOK

First published 2025

by Routledge
605 Third Avenue, New York, NY 10158

and by Routledge
4 Park Square, Milton Park, Abingdon, Oxon, OX14 4RN

Routledge is an imprint of the Taylor & Francis Group, an informa business

© 2025 Luciano Attolico

ISBN: 9781032756240 (hbk)
ISBN: 9781032756226 (pbk)
ISBN: 9781003474852 (ebk)

DOI: 10.4324/9781003474852

Typeset in ITC Garamond Std
by Apex CoVantage, LLC

To Francesca, for tirelessly supporting me and helping in the most critical moments and in important decisions, with great love and infinite patience.

To Amedeo, for illuminating every single day of our lives and reminding us in every moment that being truly present counts more than so many things we do.

Contents

Acknowledgments

The first time I thought about this book was in 2009. The idea was to help people and companies to really take care of themselves, but I didn't feel ready yet. I would like to write about my own experiences and those I had with my teams. But to write this ambitious book, I waited until 2019. Only then did I feel truly ready. But life always surprises you. I didn't know that the months ahead of me would upset many of my certainties: the Covid-19 pandemic and all the heavy consequences, the related difficulties with my company, my health problems and surgery in May 2020, the loss of my father, the negative impact of these painful events on my family.

To complete this book, I had to appeal to all my personal resources and to all the people who were close to me. I have learnt more in the last two years than in the previous ten. And I have made all these lessons I have learnt available to readers in a way I could never have imagined before.

So, my first thanks go to you, reader. I am honored to have been at your service in all the pages I have written for you and for all the people who will join us, sharing our vision of the company.

I wish they were both still with me to share the joy of this realization, but I am sure that my parents will hear well from some remote corner of heaven my deep gratitude for all that they have given me and feel strong my great love for them. The presence of my wife Francesca gave me strength and love every single day, even in moments of greatest tension. Thanks to you Francesca, much more than I can express with my words.

A special thanks goes to a person who despite his very young age, three years recently, gives me joy and love every time I look at him: my son Amedeo. He taught me that to become a good father, able to teach him how to live, I would first have to learn to listen to him, and above all, find the sacred time to be with him. I warmly thank my sister, my brothers, and the rest of my family for their constant closeness and support in this difficult year, despite the geographical distance.

If there is one person to whom I really owe a lot, it is Tommaso Massei, who has followed every step of this adventure from the very beginning and helped to improve every single page. The publication of this book is also thanks to him. Together with Emanuela Frasca, the detailed work of final revision was completed, priceless in the very short time available. Thank you very much Emanuela.

The support that Leo Tuscano has given me goes back a long way. All the topics covered have been the subject of long shared reflections over the years, and the transfer of all his "adventures" in the field with hundreds of people he has followed has been fundamental to this book. Thank you, Leo.

Special thanks go to my colleagues who contributed to the revision of the texts and the elaboration of all the stories and examples described: Gabriele Colombo, Alessandro Valdina, Morgan Aleotti, Gianluca Ferrari, Lorenzo Lucchesi, Michele Bergamaschi, Filippo Petrera, Riccardo Sivori, Giuseppe Patania, Marco Ferrante, Emanuele Schibotto, and Emanuela Frasca.

And while I was struggling with the book and my family and health problems, the company went on with great team spirit. I am grateful to Monica Tosi, Clara Faliakis, Chiara Fracassi, Luca Salvatori, Marco Cioni, Gianluigi Bielli, Gianluca Losi, Matteo Gottardi, Simone Bielli, Maria Grazia Colaianni, Riccardo Siciliani, Francesco Lischi, Danilo Pappa, Cecilia Angioletti, Leonardo Sacchetto, Francesco Terrone, Federico Visca, Rossella Di Stasi, Davide Capitanio, Riccardo Dolci, Tommaso Catalini, Guido Besana, and Jacopo Rauccio.

I can't think of Paolo Streparava as a client who trusted me and Lenovys: he is first a friend and an outstanding entrepreneur who shared a new vision of business, shaped in recent years with his contribution. Thank you, Paolo, and thanks to all your wonderful team, source of energy and discussions, even intense, but always constructive!

I think of my friend Ermanno Delogu of KUKA Roboter Italy, whom I met over 15 years ago in a specialization course for executives that we attended together when I was also working for the same company. Mutual esteem and the same vision led us to work together even many years later. Thanks to you Ermanno, thanks to Eirini Nektaria Karamani for her testimony, and thanks to all my friends at Siemens Italia for their work.

Special thanks and recognition go to Augusto Mensi for agreeing to take the reins on the project at the most difficult time. Thanks for his trust, stubbornness, and discipline in wanting to apply Lean Lifestyle principles in a complex context, together with Stefano Cantini, Luigi Lucchini, Antonio Mastrangelo, Alberto Tiraboschi, and all the rest of the Lucchini RS team.

My heartfelt thanks go to Orogel for having accepted the challenge of working with us for years to change the rules of the game in small steps, but never stopping. Starting with Giancarlo Foschi, who first had faith, and then all the other people who have pushed with their example and commitment to make things change: Lucio Fenati, Gianluca Amadori, Andrea Maldini, Giulia Rossi, and all the others—thank you. Thanks to Walter Bertin and the entire Labomar team for their relentless pursuit of excellence and human and corporate growth.

A special thanks goes to the group of friends of Sammontana for having believed in a difficult period that you can make a high-impact innovation sustainable, low cost, and in a short time, if and only if you use method, discipline, tools, and above all lots of enthusiasm and teamwork. Thanks to Leonardo Bagnoli and all his incredible team. I don't mention all of them just because I don't have any pages left!

Many thanks to Gabriele Colombo for his masterful guidance of the entire Sammontana project: he has shown great professional and above all human maturity.

Thanks to Cesare Ceraso, former general manager of Stanley Black & Decker: the friendship, esteem, and common vision that have bound us together for years lead us to make long reflections every time we talk, some of which have been incorporated into this book.

Thanks to Veronica Bini for her valuable testimony as a remote manager for over 15 years. Since her participation in one of the first editions of our Lean Leadership course in 2010, to this day she has made so much!

Thanks to Alessandro Trivillin, former CEO of Danieli and ABS—Acciaierie Bertoli Safau (steel making division of Danieli Group) and current CEO of Snaidero Cucine: your calm and respectful way of acting, but attentive to the overall vision, results, and well-being, makes you an example of Lean Lifestyle Leadership.

Thanks to my friend Fabio Camorani of Electrolux for his relentless drive to apply the Lean Lifestyle themes in the field and his testimonials that he gives every year within the Master.

A special thank goes to all the other companies that have allowed their stories and experiences with me and my Lenovys team to be told in this book: Ferretti Group, Cromology, Gualapack, Zhermack, Garavini, Marcegaglia, Fire, Gewiss, Lamec, Timenet, ELT Group, Deoflor, Akhet, NT Food, Farmalabor, Gruppo Ethos, Probios, Poste Italiane, Campari, ABS, Danieli, Stanley Black & Decker, Ondulkart, and Asonext.

I would like to thank Michael Sinocchi, Productivity Press, for the trust he has placed once again in me and his entire team for the valuable work of editing and revising this book.

My last thanks go, finally, to the professionals, doctors, and friends who have followed me and supported me in my psycho-physical recovery of the last year: Maurizio Cecchini, Debora Rasio, Davide Terranova, Giulio Zucchelli, Matteo Vicentini, Alberto Bertolotti, Francesco Lepri, Manuele Bardini, Maurizio Castellano, and Sabrina Simoni.

About the Author

LUCIANO ATTOLICO

After a Master's Degree with honors in Mechanical Engineering, Luciano started his professional career in Magneti Marelli, which is nowadays part of the FCA Group. He was responsible for the production launch of an electro-oleo dynamic system for mechanical gear shift, used by Mercedes, Ferrari, and Renault vehicles.

He moved to Siemens as Industrial Engineering Director and later as Advanced Technology Development of the Automotive Powertrain Business Unit. He was responsible for the development of production technologies and systems, as well as for the production launch of new production systems for BMW, Mercedes, Audi, and Porsche, coordinating offices and plants in the United States, Germany, France, Italy, and China.

In these industrial experiences he led many international Lean projects with mentors such as Masaaki Yutani, former Toyota sensei, Hiroshi Moriwaki, Jeffrey Liker, and John Drogosz. While working as manager, Luciano completed his specialization courses of the University of Michigan in Lean Manufacturing and Lean Product and Process Development.

Luciano is co-founder and CEO since 2009 of Lenovys, a primary European research, Lean consultancy, and training firm based in Italy that focuses on Lean Leadership and Innovation. Lenovys was recognized in 2017 by the Financial Times as one of the Europe's Fastest Growing Companies, a special ranking called "FT 1000" that includes the top one-thousand independent European Companies with the highest rate of innovation and organic growth. Together with his team, Luciano Attolico has supported relevant companies in their Lean Transformation and training projects, such as Lavazza, Sacmi, Continental, Roche, Lamborghini, Heineken, Tetra Pak, MAHLE, CNH, Frigoglass, Natuzzi, Campari, Streparava, Ideal Standard, Lucchini RS, DeWalt, Whirlpool, Leroy Merlin, and many others.

Since 2024, Lenovys has been part of the Tinexta S.p.A Group, a company listed on Euronext Star in Milan, included in the European Tech Leader index as a high-growth tech company, active in Digital Trust, Cyber Security and Business Innovation services.

Luciano has developed the Lean Lifestyle® approach, a management framework that aims to achieve the Lean Thinking as a Lifestyle. Its ultimate goal is to get more value with less stress and waste of people energy, bringing at the same time well-being and high performance in the company.

Presently, Luciano is actively engaged as a keynote speaker in the fields of Lean Leadership, Innovation, Human Energy, and Change Management, bringing his long experience, his personal philosophy focused on human value and on the continuous research to achieve more results with less efforts and higher well-being.

Luciano is author of the Italian best-selling book on Lean Innovation, *Innovazione Lean* (Hoepli, 2012); co-author with Jeffrey Liker of *The Toyota Way. I 14 principi per la rinascita del sistema industriale italiano* (Hoepli, 2014), a special Italian edition of the classic Liker's book; and editor for the Italian edition of *Toyota Way per la Lean Leadership* by Jeffrey Liker and Gary Convis (Hoepli, 2015); author of *Lean Development and Innovation, Hitting the Market with the Right Products at the Right Time* (Routledge 2019); author of *Strategia Lean Lifestyle, Lavorare e fare impresa con più risultati, agilità e benessere* (Hoepli, 2021).

Introduction: Time Is Ripe for New Ways of Working and Doing Business

Living in Strange Times

We live in times when through any social media we can communicate, simultaneously, to a potential audience of millions of people.

Thanks to these same tools, we can receive huge flows of information in a very short time, much more than we used to receive in the past and much more than our brain can process and retain in an entire lifetime.

These are times when even in our companies the amount of information processed, exchanged, and stored by internal and external communication systems, management systems, machines, and intelligent production lines has grown shockingly.

However, even though we are faced with a material productivity that has grown in all sectors, these are times when we are almost defenseless to a reduction in real human productivity, the one measured in hours worked compared to the results achieved.

These are the same times when, according to the optimistic predictions made by futurologists 40 years ago, we could have worked no more than 15 hours a week and produced many more results on an individual and corporate level.

Something did not go, according to those rosy predictions. Beyond the objective reductions in hours achieved over the years by many wage categories, we all work on average more and substantially worse. In the age of continuous connection, even when the official working hours decrease, for many people work continues in e-mail read at home and in phone calls or video calls outside office hours, effectively expanding the actual working time and crumbling the physical boundaries of the professional environment. The remote work in which the Covid-19 has catapulted us has accelerated an already ongoing process of cancellation of the boundaries between work and private life.

Our days have become like frenetic carousel rides, in which, with indecipherable variability, clashes and encounters alternate attempts at concentration and distractions, accelerations and sudden stops, and fears and emotions of all kinds: from despair to optimism, from anxiety to euphoria. Sometimes motivated, sometimes not. And all this lead us to conclude our days often with exhaustion and without a clear perception of having achieved anything noteworthy. Not necessarily noteworthy in absolute terms, but in relation to the enormous energy we feel we have lavished. This is how we find ourselves, in the evening, with the perception of not having generated the maximum possible value for ourselves, our family, and our company, and we project ourselves, tired and stunned, in the new ride of carousel of the day after.

Why is it important to talk about the highest possible value today?

I have always been attracted by the idea of intercepting people and companies who have understood and done things that I can humbly learn, lessons and principles. And in these times, the opportunities are many. This interest has been with me since I started working, now more than 30 years ago.

I confess that along the way I have always tried to find the meaning of it all. The reason. And you don't find these answers just by working, but looking for a balance, though tiring, between your busy professional life and your personal life.

In all these years of personal research, with thousands of people I have met along the way, I have internalized one thing: many of us spend a large part of our active time in companies or otherwise struggle with our work. It eliminates the time needed for all home-to-work commutes and vice versa. Eliminate the time you need to sleep, and then the time you need to wash, and then the time you need to eat. Throw in some much-needed rest time as a much-needed breath of fresh air after exhausting days or weeks at work, and you will realize that, in practice, the remaining active time that can be used for other personal projects, hobbies, sports, family, friends, social life, and anything else you can think of outside of work is a very small part of the total. You will tell me that perhaps you were expecting a slightly more exciting discovery after so many years of work. And so, I go a little deeper in sharing with you other aspects of awareness that I have gained over the years.

It is not just a question of numbers, hours worked versus total hours available, but of the consequences that the way we live those hours have on our entire existence. We talk about closely related to both the performance and the professional results we achieve, both to everything we obtain and live outside the working perimeter. I may sound cynical in some people's eyes, but I believe that almost all companies now prevail working methods that lead to heavy side effects. As Yuval Harari states in his fantastic book *Sapiens. From Animals into Gods: A Brief History of Humankind*, we live in an age of abundance, an age in which the availability of tangible and intangible goods is increasing.[1] However, something strange has happened. A paradoxical "work mule" effect from which man just cannot escape. Despite all this abundance, our life has not been simplified hand in hand; on the contrary, we do not enjoy much real well-being compared to the past, and beyond the average lifespan that has increased, new and numerous "aches and pains" and illnesses now accompany us for the long years of our maturity and our being actors in this whirlwind increase in general productivity.

And many of these distortions are precisely nested in our own companies.

Facing an external world of technologies, social and political changes, changing structures and contexts, with a level of complexity decidedly increased and in which everything revolves much faster than in the past, companies seem to insist on using outdated and hyper-reactive organizational models and behavioral modes in response.

Paradoxically, in the company we respond to the uncertainty and complexity of the outside world with greater variability and complexity.

It is like wanting to use summer tires even in the winter season and even expect to go at maximum speed, maybe cornering, on a mountain road covered with a slab of ice, without even being taken behind the snow chains!

The way we are working has become enormously complex and often does not bring maximum performance, in terms of both time and energy expended. Equally, often it does not ensure the maximum potential of people in the company, makes complicated even simple things, does not favor the generation of individual energy nor the real exploitation of talents and know-how company, does not guarantee maximum results for the company itself, and at the same time has a negative impact on people's well-being in both the short and long term. Numerous studies have shown the correlation between stress and long working hours with several diseases considered "typical" of our times: type 2 diabetes, chronic inflammation, and depression. Numerous sources have demonstrated the relationship between the loss of cognitive and cerebral abilities and the way we work decay over time of our short- and long-term memory, lexical and semantic capacities, concentration abilities, and level of physical fitness.

All in all, we are witnessing a strong growth in the forms of unease experienced daily in working environments, to which must be added the numerous states of frustration and situations of lack of professional satisfaction that reflect negatively in all other areas of our lives. But has this huge human cost, beyond the global increase in per capita income and material productivity, brought a concrete increase in value for people and companies?

We enter a slippery evaluation ground, because considerations can become subjective, but some research gives us bad news on this front as well. On average, a modern-day manager takes about six weeks a year more than 20 years ago to achieve the same professional outputs, and the development times for new products, commercial launches, and order processing have generally been reduced in all sectors, but they are not remotely comparable to the reduction of the cycle times of the productive means and to the data exchange speed of all the technological devices at our disposal. Think about the processing times of any CAD-CAM system for product and process design, think about the processing speed of modern business management systems or the speed of communication in general, and you will realize that the progress of overall business performance indicators is much lower than the local progress of individual work centers and system parts. It is like we poor human beings still do not figure out how to handle the enormous complexity into which we have fallen, and it is as if we have not yet learned how to best manage the technological potential at our disposal.

> Today we run panting after all our tasks, desperately seeking to keep up with everything, but we have a lot of work to do.
> We work harder and worse.

And what is happening in the workplace has disturbing human and social implications even outside the professional sphere. We plastic shape our poor billions of brain neurons with hyper-reactive, hyper-fragmented, and hyper-unconscious behaviors, dozens and dozens of times a day, hundreds of times a week, thousands of times a month. And we end up getting used to this way of doing things even outside of work, turning it into a way of living and a way of being.

We struggle to find concentration at home, with family, with friends, in sports, in a hobby, or on holiday. We don't know how to listen deeply to those in front of us, and we do not know how to take in the emotions of our child or partner; we basically lose clarity and find ourselves running reactive even outside of work.

Gloomy picture? Sometimes we prefer not to see reality, especially when it might make us see things we don't like.

Neuroscientists say that under particularly stressful conditions we can be affected by a syndrome quite comparable to that of Stockholm: a term coined by psychologists after a dramatic event in August 1973 when an escapee from a prison in the Swedish capital broke into a bank and held four people hostage in a very cramped space for six days. The hostages in the end feared the police more than the kidnapper himself. This psychological paradox is called Stockholm syndrome, an automatic emotional reaction to a trauma, developed on an unconscious level. The syndrome, later detected and studied all over the world since the events in Stockholm, involves a high state of psychophysical stress, which increases as the protagonists seem to accept living in a threatening environment that forces them into new situations of adaptation and the consequent regression to previous stages of personality development. This condition of forced adaptation leads us to a comparison with how often it happens to many people who, despite complaining about their lives or particularly stressful and dangerous situations, cannot easily break away from them. On the contrary, they end up being almost attracted to it, unconsciously building a deep victim-blaming bond.

Calmly observe your life and that of those around you, observe.

How long can you stay focused on one topic without distractions, without interruptions, and without thinking about the hundreds of other things you should do?

How long can you listen to a person talking to you, without thinking about anything else, paying attention only to what you are told?

Do you think your human and professional potential is as fully expressed as you could have been so far?

Do you think your team, or your company, has expressed the maximum human and professional potential?

Do you think your personal and working life is simple, fluid, light, or complicated, very complicated, or perhaps unnecessarily complicated?

This also explains the success of so many development, motivation, and personal growth paths.

It is not by chance that we witness the birth and proliferation of coaches, motivators, gurus, and would-be gurus, who bring bursts of energy and motivation to so many people, sparks of fire that flare up, but then go out as isolated positive episodes. It is as if we have a big house that is on fire, our corporate world, where we spend most of our active time with decidedly sub-optimal rules and ways of working, and we are content to solve the problem of the smoldering flames in a small area of the garden surrounding our house. It will serve little purpose to extinguish only that circumscribed area. We need to face the situation inside the walls. We cannot let it continue like this because we are putting our house and garden in danger. That is work and private life. And it will serve little purpose to hope to change jobs, change lives, become autonomous, seek personal freedom, or true wealth outside work, perhaps motivated by the "guru" on duty.

Forgive me, but I keep using the landscape metaphor. The world is made up of houses and gardens, large, with roads connecting urbanized spaces. Whatever we do, we will need a roof over our heads and green areas—shelters to protect ourselves and our families.

> I believe we all must work to evolve our companies to make them truly comparable to houses with gardens. Make them places to feel safe, where we prosper and progress to achieve great professional results and great well-being at the same time. Companies that help us live better inside and outside the company itself.

Through this book you have in your hands, I want to make my contribution to the realization of this dream. I want to put to good use my work experience, the lessons learned with thousands of people met and hundreds of company projects carried out over the past twelve years with my company's team. I believe it is possible to make companies a place where we can pursue excellent results and at the same time profound well-being for ourselves and our employees. I believe it is possible to achieve high levels of efficiency and operational agility and at the same time have people who feel fulfilled, proud, motivated, and energized at work, without having to choose one or the other. I believe it is possible to pursue corporate excellence and at the same time cultivate one's own excellence as a way of life outside the company—without compromises.

The time is ripe to break down some of the paradigms that dominate the prevailing way of working and doing business. This is a book designed for those who want to respond in a new way to the changes and daily challenges of work and business.

The book addresses, therefore, the key problem of every working person, of professionals, managers, and entrepreneurs at the helm of companies of any size: how to combine the need to produce more and more results, in less and less time, to work "better" and, at the same time, have a lifestyle that can give us physical, mental, emotional, and social well-being.

Based on research and studies of best business practices from around the world and sharing the stories and testimonies of Italian businessmen and women with whom I have worked in recent years, this book aims to provide the basis for building a new business model, a Lean Lifestyle Company. A company that aims to create maximum value and express its human, technological, and organizational full potential.

The proposed business model has its roots in Lean Thinking, the Toyota-originated philosophy that aims to increase value and reduce waste in the company and addresses the most neglected category of waste in companies, those related to people, to the point of touching the strategic governance of the company itself to combine both far-sighted vision and operational pragmatism. By evolving Lean Thinking into Lean Lifestyle and turning it into a strategic weapon at the service of the company, this change of focus is brought about by a simple consideration. In contrast to the historical period in which Lean Thinking was born, work in our companies has become increasingly "intellectual" and is mainly done in the office or from home, with a computer, tablet, or smartphone. A lot has been done in the manufacturing world in recent years, but very little has been done in the way people work, despite the entry of technology into all areas of business. By eliminating waste in the way people work and enhancing the potential of human assets, we will get more valued, more effective and efficient, more satisfied, and more motivated people, with more well-being. As a result, companies will find a new harmony that fosters and accelerates overall growth and increases performance.

In the book you will find numerous examples, tools, and step-by-step methodologies that will help you start applying the operating principles of the Lean Lifestyle yourself to achieve benefits for yourself, the people you work with, and the performance of your company. I hope this book will help raise awareness that every choice we make always has an impact.

Through the evolution of the way we work and do business, we can all leave a tangible mark that starts at the company boundaries, touches ourselves and the people we work with, and even positively influences society and the environment where we live.

Enjoy your reading and, above all, have a good journey in building your Lean Lifestyle and your Lean Lifestyle Company.

Luciano Attolico

Note

1. If we look at the numbers involved, we find that in 1500 the average per capita income was $500 and in 2016 the same number reached $10,739. The global world production, instead, in 1500 was $250 billion and has become approximately $76 billion billions in 2016. Data source: World Bank Group.

Chapter 1

Traveling to My Lean Lifestyle

When You Think You Have Hit Rock Bottom

April 2020—the Covid-19 pandemic has erupted worldwide—fears never felt before, disorientation, and the inability to decipher the present and the future.

When are we going to live and work in full normality? When can we simply shake hands and hug each other without fear of being infected? When do we get back to travel peacefully? Is our health, our very life, at risk?

These were just some of the questions that occupied our minds in those days and that still haunt us today, February 2021, without any definitive answers. To prevent that negative thoughts took over, I tried to remain lucid, to be close to my loved ones, and to manage the company situation, which was anything but rosy: protecting all the Lenovys colleagues and their families from the contagion, managing the consequences of the drastic drop in work and the difficulties in finding alternative forms of profitability, and the painful decision to reduce all our salaries. And then there was still to continue customer support and activities on those few projects that had not been interrupted by the sudden closure of companies.

I tried to focus my energies on what I could control or modify, without being mentally devastated by what was outside my control area.

But, when I thought the worst had already been achieved, I realized that everything is always darn relative. On May 5, 2020, I found myself, incredulous, in an icy operating room with no control of anything. The caring attention of the nurses preparing me for surgery, the papers for the anesthesia, the ritual explanations, and then the memory that the only thing I could and should do was to remain calm and still and breathe or hold my breath when the vascular surgeon asked me to. As I watched

DOI: 10.4324/9781003474852-1

the ceiling lights, almost hypnotized, the vascular surgeon first conducted a long diagnostic examination called "coronary" to quantify the location and characteristics of the damage suffered, and then performed a delicate coronary angioplasty.

I had had a heart attack. A part of my heart was now in a state of severe necrosis because it was no longer sprayed with blood from the artery that had become completely obstructed. A few more hours, a few more days, and I would have said goodbye to this earth experience forever. The operation was complicated by the fact that I had had a heart attack a few days before, on April 30, at the end of a training session. I had exchanged the symptoms—severe chest pain, deep fatigue, cold sweating, jaw pain, nausea—for a particular state of fatigue caused by the heavy session done. I was preparing for a CrossFit competition, the sport I practiced intensively and with great passion in the past five years preceding the heart attack. The last thing I could have remotely imagined was that I might have a heart attack. It was a bolt out of the blue. No warning signal beforehand, blood tests almost always in the normal range, no smoking, no hypertension, regular sport activity, ideal weight, attention to nutrition, and above all I was strong of the certification for the practice of competitive sport that I had obtained every year without problems (including the one obtained a few months before the heart attack!). The only clue that could make me think was my father's familiarity with heart attacks and strokes: even in his case a serious unexpected event had materialized when he was just over 50.

Perhaps due to the Covid situation and the fear of entering a hospital emergency room, my family doctor, contacted twice following the symptoms I described, had unfortunately confirmed my thesis (seriously wrong): posthumous of a too heavy workout. The treatment of that afternoon, which I will never forget, was an anti-inflammatory and two hours of sleep under the covers. I could have never woken up again. Instead, I woke up two hours later and resumed my activities; a little dizzy, but all in all happy to no longer feel nauseous and to have appetite again. I had dinner and started working again. Unfortunately, what had happened was that my heart had stopped asking for help and had also stopped loudly communicating those symptoms to me. The process of "death of the heart tissue" had started because of the ischemia suffered. If you intervene with due urgency following a heart attack, two things happen: first, it is very likely that you have saved your life, and immediately after, you can restore the situation of that part of the heart muscle that is no longer watered of blood and oxygen, because timely intervention brings blood to circulate back in the vessels that have clogged. I, on the other hand, unknowingly started playing Russian roulette with my life for every extra minute I spent outside a hospital in that condition. And it had been a long time since April 30 until the afternoon of May 4. In between, there were several races with my son Amedeo and a training session, less heavy than usual, but still very high-risk. On May 4, pushed by my wife and my colleagues, Tommaso, *in primis*, who gave me the precious contact, I phoned, without even being so convinced, to a cardiologist of Pisa, Dr. Maurizio Cecchini, to whom I described the symptoms I had. He ordered me to join him for a checkup. You should have seen his face at the sight of the electrocardiogram first and the echocardiogram later. "You are, still, alive, by miracle," he scanned, looking me in the eye. Then he asked for my wife's phone number and called her instantly to explain the gravity of the situation.

I owe him my life. He didn't waste a minute more and sent me straight to the emergency room. From that moment on, my adventure in hospital began, forgetting my fear of Covid contagion in the face of an even higher and more imminent risk. After setting foot in the emergency room and seeing how critical the situation was, I was no longer allowed to get off the stretcher or out of bed, not even to go to the bathroom. Containment, preparation, and execution of the intervention and then the stay in the coronary intensive care unit for several days until the risk of death disappeared. When I resigned from the hospital, I was reassured that I was "out of danger," but at the same time informed that I now had to live with a significant reduction in my heart function. After a few days, I gradually started practicing my new sport: long, peaceful, and "healthy" walks on the promenade.

I won't hide from you that in the endless hours spent alone in the hospital, thousands of thoughts began to crowd my mind: What do I do now? What changes in my life? Who am I now? Am I no longer the father and husband I used to be? Suddenly, I am no longer the seasoned athlete and sportsman, the visionary entrepreneur, the consultant, and the passionate trainer of a few days ago. At first, I experienced everything that had happened to me in a very negative way. I saw limitations and sadness everywhere. But then I realized I was living a fantastic opportunity, a kind of a point of no return. For many years, I had worked toward the construction of a business model based on Lean Lifestyle principles, which would transform Lean Thinking—more value and less waste—into a way of life inside and outside the company to achieve deep well-being and more results at the same time, through new ways of working and organizational setting. But in those moments, I began to understand, with my belly and my soul, the impact on myself of what I used to decode more on an intellectual level. I began to see everything around me with different eyes. I felt happiness and gratitude for the mere fact that I was still alive and, after all, without any particular disabling restrictions. I was only "obliged" to change my lifestyle once again, going even deeper than I had already done over the years. I started to see what I could do again and differently than before. Hugging my wife Francesca and my son Amedeo suddenly had a different flavor. A new awareness was taking hold: that of the importance of living each moment of one's life with intensity, as if it were the last. Because none of us, unfortunately or fortunately, knows when it will actually be our last. A different kind of intensity than before, however, because the focus was shifting from continuously doing what it takes to achieve professional and personal goals to feeling more and more yourself before diving into action. I understood even more the meaning of presence toward oneself and others. I understood with my heart and not just with my head what it means to be close to another person, what it means to empathize with the other person, and what it means to understand their emotions and feelings. It is not that I was not empathetic and present before. I have spent a lifetime working on myself to be more and more so, but the most important thing I've realized is that if you are not used to be really close, attentive, caring, and compassionate with yourself, it is much harder to be really close to others. I started asking myself the same questions as always with a new spirit and found different answers than before. What does well-being really mean? What does achievement really mean? What is, really, success for a person? How does all this find space and meaning in the organizational sphere, in the life of men and women who work and lead companies every day?

Succeeding

The corporate world is the world of doing and results par excellence. Goals and all kinds of actions to achieve them, in one way or another. I started by reflecting on my company, Lenovys, and I wanted to extend my considerations to my whole team. What was our point? How could we make individual missions meet company missions? We realized that the difficulties of the Covid period had paradoxically united us even more and strengthened us as a group. **The first great desire I felt and shared with my colleagues is that Lenovys, after years of work on Lean Lifestyle, would be able to turn it into a real business model** and become an example and a sounding board itself. First, we had to act on our own skin and then turn the focus to our customers. We have thus started, in advance, the path that gave rise to the new strategic plan for the period 2021–2025. This was not the first time we worked out the multi-year strategic plan. Since 2009 it has been an annual appointment that we have never missed, but it was the first time that the company strategy became consistent with my personal strategy and that of my closest colleagues. It was the first time we had devised a strategic plan focused on a few, but vital, driving directions for us and the market.

If I wanted to evolve my lifestyle on a personal level, I also had to evolve the Lenovys management model. Getting more results and more well-being, from myself and from the company I have the honor to lead, has meant reducing the sources of unnecessary stress arising from too many open fronts and focusing on a few key elements. It has meant shaping a more agile and flexible organization, with even wider spaces of delegation and autonomy, in which people enhance their strengths and passions with a great sense of responsibility. It has meant consolidating organizational cadences for the entire new year through which we could align ourselves and find countermeasures to deviations from the single strategic plan of the entire company. In this way, I was able to find time and energy to complete the book you are reading and at the same time devote the right amount of time to other business and family responsibilities.

Taking a look from the inside out, the first new strategic direction was to draw up a manifesto for the Lean Lifestyle Company, making it reference and inspiration for our work with customers, for our communication, for our behaviors, for our professional choices, and for anyone who wants to freely adhere and use it. You will find it in the last chapter of this book.

With this personal and business process, I was able to answer the question: what does it mean to be successful?

The form of success I wish for myself and others is first of all to give the chance to dream and then to have the strength to realize one's dreams, making work and business not an end in itself but a formidable tool to realize dreams that go beyond the personal sphere.

My dream is to turn as many companies as possible into generators of prosperity and well-being, to turn companies into places where it is worth going every day—places where the dreams of entrepreneurs, managers, and workers are created and made with passion and lightness, places where people's talent developed, preparing them not only for professional excellence but also for excellence in life. I have spent many years trying to achieve something and struggling to get it, but now I am beginning to enjoy the fact that this is the kind of success that fills me with joy. Every day I take one more step in this direction, and I no longer must wait to reach who knows what to stop and enjoy the fruit of my efforts. The journey itself, in this way, is a source of pleasure and well-being, with many small stops to reflect and enjoy the road you are on. One, two, three, four, ten, a hundred, or a thousand, no matter how many companies will be inspired and oriented to Lean Lifestyle, and no matter to what extent, no one will measure the size of the dream of every company and every entrepreneur. What matters is the direction one gives to one's company and the process one puts in place to walk in that direction, without ever losing sight of the meaning of the company's existence.

I think about my current life with more space for me, for my family, and to take care of my body and my mind. I think about this book that is finally taking shape, and I think of my dreams and inevitably my thoughts go to my company. One by one I look into the eyes of my colleagues, the friends who together with me have built Lenovys, and all the others who have joined in over time. In each of them I see a person I respect, appreciate, and thank. A person who has chosen to contribute, with their uniqueness and experience and with their skills and daily work, to the continuous evolution of a company where every day is worth going.

I have a really rewarding awareness: I am enjoying a journey, made together with my family, my team, my friends, and the clients I have the honor of working with. The thought of moving on with what I'm already doing, maybe improving something every day, fills me with joy.

That's what success is for me and that's the kind of success I wish for everyone.

Decisions and Habits

I think that the life of all of us is strongly characterized by two great thrusts: the decisions we make and the habits we consolidate along the way. The first defines the direction of the path, and the second the way we will walk along this path. The sum of the two thrusts shapes who we are today and who we will be tomorrow. Let us think for a moment, what are the few decisions that have brought us this far? The city or country we live in, the course of our studies, the person we have at home, the profession we have, the company we work for, or we have founded. In the end, nobody forced us to choose. It was we who made most of the choices that gave direction to our lives.

What about habits? **About 70% of everything we do every day is simply the result of habits acquired over time. Small or large doesn't matter**. The sum of these habits has made us who we are, along the chosen path. Some examples? Do you play sports every day or every other day? Do you have the habit of keeping your e-mail and message alerts switched on? What do you usually eat? Do you make a habit of meeting your co-workers on fixed days every week or when it happens? Do you make a habit of stopping every weekend for an hour or two to reflect on what went well and what can be improved to plan well for the coming week and to decide what to do and what not to do to move in your direction? Or do you try to cope without specific preparation every new week with all the river of activities coming up, simply trying not to miss even a shot?

A person can make decisions that lead them to change direction and acquire habits through work on themselves, discovering strong internal motivations for change, or they can be forced to make choices and change because "shocks" have come at them from the outside.

I experienced both moments and used them to walk in the direction of my Lean Lifestyle. The last jolt for me was the heart attack. So strong that I might not have been able to get up from the ground. Other shocks were less violent, but perhaps just as profound and very important in making me who I am today.

So, I continue to keep the curtain of my life open to share with you some more lessons learned over the years.

Working Mules

"Tell the cook you get compliments from the room. The spaghetti with seafood was fantastic." Throwing dirty dishes on the counter and taking other freshly prepared dish to take to the dining room, the waiter had shouted that phrase, as if it were a normal thing, among other things that he had said. I was on the moon: that cook was me. I had taken advantage of the real cook's temporary absence to process an order in his place. I had been working in the kitchens of that restaurant for months, washing dishes, glasses, pots, preparing hors d'oeuvres and fruit, but I had never been allowed to put my hands to the cooker. I was 13 years old and perhaps their choice was understandable. I have always worked hard, ever since I was a child. I am the last born in the family. I have two brothers and a sister. My father was a factory worker, my mother was a housewife. Their level of schooling was quite low. My father finished school in the fifth grade. My mother did not finish that either. One value, however, was passed on to me so strongly that I felt it on my body for years—the value of hard work and study. My father worked from 8 a.m. to 4.30 p.m. at a Fiat dealership and from 5 p.m. to 8.30 p.m. in a body shop. He dismantled and reassembled cars and repaired them with a skill recognized by everyone. I forgot, he also worked on Saturdays. Until he was disabled at the age of 54 by cerebral stroke. Then he got back on his feet but continued to work only partially. All the sons in the family started working at a very young age. That was the rule. Family of working mules. You had no other choice. My mitigating factor in working hard all year round was that I could attend school in the winter months, because I always had high grades and I had never failed. So, I was only allowed to work in the summer, during the Christmas and Easter holidays. Not bad, right? When someone complained at school because they were bored or tired, I would laugh. I thought about what he could say if he had spent a few days with me from 9 a.m. to midnight working in a restaurant, except for a break, from 4 to 7 p.m., where I went to play five-a-side football and have fun with friends on the beach. My mother, on the other hand, always made sure, every single day during school time, that I studied. She didn't know what I read or what I wrote, but she cared that I did. I had many jobs, between the ages of 10 and 25. Always alternating with studies. Dishwasher, waiter, butcher, cook's assistant, bricklayer, grape picker, and sandwich maker. The work mule syndrome took hold of me so much that I carried it with me, even when my living conditions changed. I have always tried to get rid of it, but it has been a long road and perhaps the process of liberation is still ongoing. I have always been looking for improvement, to work less and have more results with more well-being. Only I confused well-being for years with the overflowing energy I always put into everything I did, work, sport, friends, family, and fun. It is important to have energy, but sometimes we confuse movement, adrenaline, motivation, and inner charge with what could really give us well-being and prosperity. I think I still have to come to terms with my working mule syndrome from my own experience and history, but each of you will surely have your own story and your own accounts to settle.

What is preventing you from working less and better and above all from seeking your deeper well-being?

Always Ready to Find Excuses

"What are those pieces down there on the floor?" he asked in barely spoken English, with a strong, drumming Japanese accent. His body movement was swift, unhesitating, and he stood with the finger of his right hand outstretched to point at the pieces, waiting for an answer. I was an intern in the Bari plant of Magneti Marelli, 1996. Masaaki Yutani was a former Toyota manager called directly by the then CEO Domenico Bordone as a special consultant to transform the group's factories according to Lean Production principles. At the end of each visit, he assigned tasks and improvement actions that he would promptly verify during the next visit. His methods were far from gentle, but very incisive.

The tentative answer to his question was: "The cleaning service hasn't passed yet, that's why they are still there." His counter-response pierced me like a blade: "Always ready to find excuses!" It's not that he was particularly nervous during that visit. His manner was always so brusque, almost jerky. "Tell me what parts they are and how much they cost!" he exclaimed. We explained that it was the anchor, a small steel tube that fits inside the petrol electro-injector, one of the plant's flagship products. And we also told him the exact cost of the part. We continued our visit and, unfortunately, there were a couple more pit-stops for more parts on the floor: more scenes and more explanations about parts and relative cost. Until something happened that has left me an indelible memory over the years. "Give me money of the same value as those pieces," he said. Incredulous about the strange request and curious about what he would do, we gave him the money. He quickly went to collect the pieces scattered on the floor and replaced them with the money he had received. He asked us to go away with him for a while and to watch, hidden, what would happen. A worker, a few minutes later, stopped and bent down to collect the money on the floor. Plot twist. Masaaki Yutani jumped like a tiger a few centimeters from that worker's face and ranted in an English that only a few understood verbatim, but which they all understood in substance: "Why money yes and pieces no? They have the same value!" This lesson has marked a path in me. How many times had I passed through those production lines? How many times had I already seen those pieces on the floor? Maybe I thought about doing or saying something. Then came the next day and yet another. And I thought again about doing something. I must have gone over and over it. Until I stopped seeing them. I stopped paying attention to them anymore.

> In company and in our lives, we do not realize the waste around us because we are addicted to it. We no longer notice them, we no longer see them, but they continue to be there.

Since my time at Magneti Marelli, I have always tried to keep this in mind, both in my personal and professional life.

> And what waste do you no longer see in your work and private life?

The other big lesson learned during the sharp visit described was the concept of value. Money and components have the same value for some, but not for others.

> If there is no alignment in the company on what value is and what it is not, it will be very difficult to move compactly in the same direction towards value creation and continuous improvement.

I fell in love with the Lean principles from the very beginning of my professional adventure also thanks to these lessons in the field. More value and less waste seemed to be in line with my personal and professional quest. It was in this area that I had to focus my efforts: Lean Thinking in companies and in life. My thesis was completely oriented to the application of those principles to the processes and organization of the plant. I was hired in 1996, immediately after graduation, and continued to work on the experiments launched during the thesis internship, for about a couple of years, reaping results—we reduced the materials circulating in the factory by about 40% and halved the throughput time in the factory—and personal satisfactions.

Problems Always Go Looking for Those Who Already Have So Many

> *"Do you want to be the first in logistics or the last of the process engineers?"*

This was the dry and direct question that my possible new boss asked me, in his strange attempt to convince me to change my position. In 1997, Magneti Marelli's Bari plant was awarded the "Selespeed" project, derived from the Formula 1 experience, which allowed for the robotization of the mechanical gearbox. To become a process engineer, accepting the invitation to join the

Selespeed project, or to continue a secure career in the logistic sector where it was spoken about my possible move as head of the entire department?

There were many unknowns in accepting the new project. It was a new product that could also turn out to be a flop. Besides, it was a new job for me, and I knew nothing about process engineering. After all, difficulties and challenges have always attracted me and I was also intrigued by the talks I was hearing from all the engineers dealing with technologies, production systems, timing and methods, and assembly line design. There was a world of skills that I didn't have—at least in the field. My degree in mechanical engineering, with a focus on plants and production, had made me study many related things, but in practice I had seen very little of them. So, I chose to become the last of the process engineers and move toward the new. It was three hard but wonderful years in which I was able to apply Lean principles to the area of new product development and new production processes. But perhaps the most important thing I learned was at the end of a long day, struggling with one of the many quality problems that we used to have in that start-up period. It was late, well past the normal exit time, and I hadn't realized that everyone had left. I was just grappling with the results of some lab tests and analysis to understand the nature of some problems we were having. Only the distant lights of the production lines that were finishing their second daily shift kept me company. Constantine, my new boss, entered the office, a large open space with glass walls separating us from production. He came and sat right next to me. He stayed beside me for a few minutes trying to figure out what I was doing. Then, I broke the silence with a question: "Why do I have such difficult problems here?" His answer was memorable:

"New problems always go to those who already have so many, because then they are much more likely to be solved. Mind you Luciano, those who never have problems are poorly trained to face them and solve them. And so, their desks will remain empty. Yours, on the other hand, will always remain full, unfortunately for you, but fortunately for the problems".

He stayed by my side the whole evening and helped me. We went home tired but proud to have overcome one more obstacle. And I got an important lesson.

A leader teaches more by example than by many words. If you want to help someone grow, it is important to be next to him on his playing field or on the *gemba*, as the Japanese say.

The simple, but powerful rule of *genchi genbutsu*, go and touch with your hand where things are happening, would always lead me to spend a lot of time outside the office, in the midst of the problems of the line workers and my colleagues, with customers or with suppliers. However, the real lesson on this subject could only come from Masaaki Yutani. On one of his visits, he stopped in front of a work center that looked clean and tidy in appearance. He observed and said nothing. But this did not surprise us, because for him it was normal to stand still and observe the movements of the operators, machines stopped or being worked on, and materials stopped or in transit from one part to another. He could stand still for a very long time to observe, and even then, this detail might have seemed peculiar, for us Westerners who are used to running without ever stopping too much on things. "Have you cleaned everything here?" he asked. The question was not accidental, because in that same work area on the previous visit he had ranted about the lack of order and cleanliness. The answer in chorus was affirmative. And it looked clean and tidy. Faced with this reply, we saw Yutani bending over and rubbing his sleeve under the machine. He got up with an embarrassing oil slick on his jacket that left us all livid with shame. He spoke little on that occasion. But he taught a lot too.

> How much time do you spend sitting at your desk, in front of your computer or with your head bent over your smartphone, and how much time do you spend on the field, in the middle of real problems?
> How much time do you spend with your co-workers, colleagues, suppliers, and customers?
> Are the relationships you are developing real and solid or just hierarchical or virtual?

I try to ask myself these questions every day, because especially in this chilling pandemic period, the risk of isolating oneself and closing oneself off has, unfortunately, increased dramatically.

Go and see for yourself. Genchi genbutsu.

Prepare Your Resignation, Because Tomorrow You Will Have to Resign Them

The tension was very high. The Porsche technicians were staring coldly at the electronic display board showing the quantity of good parts produced at the end of the assembly and testing line, and unfortunately the numbers did not stand in our favor. The efforts of more than two years of work were about to fail. It was called "Run@Rate," the audit that the Porsche delegation was conducting in the Pisa plant of Siemens VDO Automotive. I moved to Tuscany, to Livorno, in February 2003, after a period in Milan in a consultancy company that lasted just over two years. At Siemens, I oversaw process engineering and production for a new family of products, the so-called high-pressure injectors. That audit, if passed with flying colors, could have given us permission to become official Porsche suppliers for the first time in history. The strict test we were subjected to would have verified the ability to produce the right quantities of finished product in the available work shifts, in the expected quality, and with the declared line efficiency targets. Additional difficulty was that the products to be assembled and tested were of two different types and for two different customer engines. So, there was also the uncertainty of a line type change from one product to the other. We had so many problems in the previous months and many of them were still being solved. That injector had never been designed, developed, or manufactured before. Materials, tolerances, couplings, and welds were all elements with heavy unknowns because the operating pressures were about two orders of magnitude higher than those of my colleagues in Pisa, world experts in the design and production of so-called low-pressure injectors for the automotive market, had been used to for years. The assembly line was designed and built, station by station, for the first time with new criteria compared to the past: from production with high volumes and few variants to flexible production with low volumes that could potentially be increased, with several possible product variants. Result, so far, was large amounts of waste and ridiculous production line efficiencies. And that audit confirmed the difficulties we were unfortunately used to. We had already burnt up a lot of our time, we had not completed the production of the batch of the first product variant required, and the production of the second product variant had not yet started. The line was a long U-shaped snake with 21 interconnected assembly and testing centers, within each of which there were several sub-workstations. In that teeming bustle of activity, I notice that Liborio, my foreman, was getting ready for the fateful type of change. I approached him and asked him what he was doing and why.

I prepare the line for the product type change. Alberto ordered me to do so.

Alberto was the plant manager. A historical figure, feared by all. At the same time authoritarian and authoritative. His career had started many years earlier, from his entry as a simple factory worker. His skills, technical, but above all managerial and relational, had propelled him to his position of prominence, which he firmly maintained for a long time, thanks also to the numerous successes and achievements of excellence that he had attained over time. Therefore, it was difficult to oppose Alberto's order.

"Have you verified the data? With the current waste and efficiency rate, do you think we can complete the first required batch before moving on to the second?" I urged Liborio. He looked at me almost lost and I realized that he had not. His eyes only told me that his goal at that time was to carry out the order he had received from his boss's boss. Stop. Nothing else. Courses and historical recurrences: I put myself at his side and we quickly began to make those calculations, from which emerged the need to wait another hour before we could do the line setup. My counter-order was therefore to wait another hour and complete a setting that would gain us speed later, based on the data available. Realizing what was happening, Alberto rushed to me asking how I had dared to contradict his order. I explained my reasons. He was furious, but at the same time extremely tense about the very real possibility of a failure that would certainly have a strong negative echo in the company's German headquarters.

> Prepare your resignation, because tomorrow you will have to resign in case things go wrong. The responsibility now is completely yours.

We waited for that extra hour, completed those adjustments, and managed to reach the tightrope to deliver both batches on time, in the expected quantity and quality. Subsequent positive tests carried out by Porsche directly on the cars would sanction a historic milestone achieved by the Pisa plant.

The next day, Alberto called me into his office.

> No one in so many years has ever dared to say no to me like that. Thank you. You had guts and head, and you took full responsibility.

Years later, I remembered this episode, not so much for the personal pride I still feel, but for the awareness that the biggest lesson learned is that transmitted to my co-workers and colleagues in those circumstances.

You must:

- Stop to observe, analyze data, apply the principles of problem solving before making decisions, even at critical moments.
- Take responsibility for your choices and do not perform tasks uncritically.
- Stay on the field with your team until the end.

But Alberto gave me the most important leadership lesson: a leader allows his co-worker to exercise, even in a critical situation, his own sense of responsibility and the freedom to carry out his choices, in the face of solid motivations.

Thank you, Alberto: it was a lesson for me and for my whole team. A wonderful team where I got to know Lorenzo Lucchesi, the one who years later would become a great friend, partner, and now the general manager of Lenovys. The many years of discussions and open confrontations consolidated between us enormous trust and mutual esteem, both human and professional.

Are you developing a sense of responsibility in your employees?
Do you train them to make their own choices?
Do you train them to solve problems methodically and rigorously, with their head before their gut?

The better we do this, the more we develop our leadership, but above all that of our collaborators.

Why Do You Do It?

I was about to make a momentous decision for me. Abandon my managerial career in the German multinational to dive into a new adventure as a freelance consultant. But I had many doubts and felt torn by a real change of professional identity. Between 1980 and 2007 my multifaceted professional experience had always been as an employee—from a small restaurant to a large multinational company. Increasing levels of autonomy and responsibility, but always with the parachute of a salary at the end of the month, health insurance, and paid pension contributions. Always covered by much broader shoulders than mine, those of the company I worked for. The idea of changing and finding myself alone to earn a living, on one hand, attracted me, and, on the other, it terrified me.

It was a sunny September afternoon in 2007 when I asked Alberto to meet me outside the company to talk. Over the years, the relationship between Alberto and I had become very close and characterized by a deep and mutual esteem. We had become friends, even before work colleagues. The traditional boss–collaborator relationship, an obsolete definition for me, but very much in use in company jargon, had been outdated for quite a while. The same thing happened several times over the years between me and various colleagues I met along my journey.

If I had made my decision to leave the company, the first to know would have been him. I raised my doubts about the choice I was about to make.

"Why are you doing this?" was his dry question. A few words, as the best leaders can do.

I spent some time searching for the right answer. It was not to him that I had to answer, but to myself.

I had a dream and I wanted to realize it. I wanted to cultivate my passions and make them available to as many people and companies as possible. I wanted to help change the way people work and do business to improve the quality of life within companies, where people spend most of their active time. I wanted to take Lean Thinking to a new level of application, closer to people and to areas where it had not yet been applied. I wanted to show that better results can be achieved by working less and better. I didn't really have it in my head how to do this, but I wanted to get started and start doing something. I felt the time had come. I explained these things to Alberto, and I don't even know if it was at this level of definition. I remember well, however, his counter-reply.

"You are as brave as you are crazy. If this is what you want, you must go. I will lose one of my most valuable collaborators, but I want to help you realize your dreams." Alberto showed a shocking and unexpected openness, giving me another lesson in leadership.

> Instead of holding someone back at all costs, try to understand
> what their dreams are and help them come true. If these dreams
> coincide with yours, you will build great things together and go far.
> If dreams diverge, the sooner you realize this, the better for both
> of you. True relationships, first. Then the issue of mere work.

The next day I resigned. In January 2008, I started my new professional life, and my first consulting project was with Siemens VDO Automotive, which in that year was divested from Siemens and fully acquired by another

German giant, Continental. It was a rather technical project, but approached with innovative Lean Development methods, which I would describe in my first book published in English in 2019, *Lean Development and Innovation, Hitting the Market with the Right Products at the Right Time* (Routledge).

How Will We Pay His Salary?

What I imagined as a lonely walk immediately turned out to be a journey made together with many friends who would join along the way. The first was Gianluigi Bielli. I had met him as a consultant during my time at Siemens. Then we met again on a project we did together at Merloni Termosanitari, now Ariston Thermo Group, thanks to the consulting company for which we both worked as freelancers. A man of great experience and passion, he taught me a lot on a human and professional level. It was he who proposed I work for our first big joint customer, Laika, with whom we would have worked for many years: still today I still thank Franco Barducci for the initial trust he had in Gianluigi and me. Franco was then managing director of the German-owned company, but with strong roots in the heart of Tuscany. Today, he is managing director at 2C Coveri, an effervescent Florentine company producing components and accessories for the fashion world.

Soon afterwards, in the same year, we gained the trust of Pietro Cassani, then general manager of Sacmi, who became our second other historical client. In a few months I immediately realized that I could not continue alone. The new job was beautiful, with adrenaline, but absorbed me more than before, with much more travel and commitments that required immediate answers and study, lots of study. And I realized that the dream needed investment, people, and structure.

A company was to be born and we called it Lenovys: the happy event is dated January 21, 2009, with the registered office in our accountant's office and an operational office in a flexible workspace in Milan. My first meeting with my friend Gianluigi, who had become a co-founded partner, was exciting. We prepared the fledgling company's first multi-year plan, applying the method that you will see explained in Chapter 13. After all, we did not really know what we were getting into. At that time, we only had three customers and none of us had ever sold anything. In the first three sales, we had simply been bought by the respective clients who knew us and trusted our technical and management skills. We started hiring administrative staff and consultants to reinforce the team right away, but soon we realized that we lacked high-profile figures to sustain growth. We decided to bring in Leo Tuscano, an old friend of mine from the years of the Milan interlude in 2001–2002, when we worked in the same consultancy firm. He was the one who interviewed me for the first time in 2000, together with another brilliant consultant, Matteo Gottardi, who would join Lenovys a few years later. I had traveled with Leo several times to the United States, for a big project he oversaw. He was very bright, able to speak like a native English speaker, with excellent processing and analytical skills, great listening ability, and able to range from hyper-technical to organizational and behavioral topics. Above all, we were bound by a deep friendship cultivated over the years. But there was one small problem. Leo had made a career and a lot of it. He was, in 2009, a well-paid executive at an international consulting firm. He was the father of two beautiful little girls and was now rooted with his family in Milan. We, on the other hand, were a tiny start-up with lots of enthusiasm, big dreams, and few resources, because we had no external sources of liquidity or financing partners and every euro we earned was always re-invested in the company.

"How are we going to pay his salary?" I asked Gianluigi incredulously, when Leo told us that he had accepted our proposal to work with us. His answer was the same one he would give me on many other similar occasions: "If we have people to pay every month, you will see that we will be forced to find new customers and new jobs to follow and have followed. We will find them and so we will pay their salaries and ours."

So, it was. Leo was the company's first manager. He had accepted because he was attracted by our vision, our dreams, and our enthusiasm. But above all, because he trusted us. And we felt obliged to return the same level of trust. These first steps in my new adventure as a small businessman, along with other events of a completely opposite sign, taught me a great lesson.

In business, as in life, the first ingredient to seek is mutual trust. Any business venture is always a team game. If you don't create the right team, with the right people, you will not get very far. And the right people are not necessarily the most competent ones. But those in whom you can place the greatest trust.

Skills can be created. But it is only with trust that the bonds created will be so strong that it is difficult to break them at the first gust of adverse wind. With mutual trust, the climate remains collaborative even in the most heated moments of confrontation. I have made big blunders whenever I have placed trust in the wrong people, dazzled by the resources they had and I did not, lured by skills they had, and I did not. Or simply because I did not take the time to fully evaluate who I had in front of me. A selection interview is not enough to be able to assess all the elements needed to build mutual trust. You strive to make the best possible choice with the information you have and the most appropriate evaluation process. But trust grows with time and real life.

How can we not remember the spring of 2013 when Lorenzo Lucchesi, my former collaborator at Siemens, agreed to join us leaving a prestigious management position at a multinational company. At that time, he was working in China: he met me in Livorno, we talked, and he decided that he would quit the next day.

So, he did and without hesitation took to the field wearing the Lenovys jersey. At the end of the year, we realized that we had not found a minute to draw and sign the contract! Trust. Today, he is a partner and general manager with full powers as CFO and HR director.

Trust is also the word that binds me to Tommaso Massei. With him I designed the publication of my book and with him I built the company's entire communication and marketing structure from scratch from 2011 to today. One summer afternoon I invited him for a swim in a bathing establishment a few kilometers from Livorno, the city where I live: I presented him my vision and the strategic project of Lenovys and asked him to come on board exclusively. He was a freelancer and worked for companies, public bodies, and newspapers. He took a few weeks and broke off all his existing contracts one after the other.

We signed the contract after a few months there too!

I think of all the other colleagues who over the years have joined us in this long adventure. Individual trust has turned into corporate trust. And we saw this in 2020, when we all voluntarily cut our salaries during the most critical months of the pandemic. In the toughest period of that year, that of my hospitalization and subsequent convalescence, the support of my colleagues and their level of autonomy and responsibility were touching. The year ended on a positive note thanks to the tremendous efforts of the entire team, after the long months of sacrifice and fear.

From 2009 to 2019 we grew for eleven years in a row. In 2020, we contained the damage and for the first time in our history we did not grow in terms of turnover, but perhaps that is what made us grow the most from within, preparing us for new adventures. Overall, we have had the honor of serving more than 500 companies and training more than 30,000 people. This has allowed us to achieve important recognitions including the inclusion in the special 2017 FT 1000 ranking, i.e. the list of 1000 European companies with the highest organic growth rate from 2012–2016, compiled by the Financial Times and Statista. Other recognition came in 2018 from Repubblica—Affari e Finanza, which included us in the list of the 300 Italian companies with the highest organic growth rate, which was also joined by Sole 24 Ore's, which included us in the Italian list of "Growth Leaders" companies in the years 2019 and 2020, for the growth rates recorded in the years 2014–2019.

I do not believe that we can really distinguish between the professional and personal spheres, as we are one multiform container of experiences and experiences that contaminate each other. In this way, we give life to what the outside world perceives of us. We achieve our results as the fruit of the one person that we are.

The Lean Lifestyle journey is at the same time a journey of people and companies, toward a real increase of prosperity and well-being in which the economic and financial factor represents only a means and never an end.

This is why I think that the whole of society could benefit enormously if companies systematically started to orient their people toward excellence as a lifestyle, inside and outside the company.

Chapter 2

From Lean Thinking to Lean Lifestyle

Founded in 1937, the Toyota car manufacturer is preparing to freshly pass its first 90 years of life and is looking forward to its first hundred years.

Despite its more than 330,000 employees, and the more than 10 million vehicles produced each year and over 100 million vehicles sold in its history, Toyota continues to be as agile and visionary as a start-up.

And it continues to "set the standard" thanks to what is known around the world as the "Toyota method."

A method that revolutionized the way a company is managed.

From being a company on the verge of bankruptcy in the 1950s, the Japanese manufacturer has strung together the longest period of profit improvements in industrial history—from 1957 to 2007—to top the rankings of the world's car manufacturers, in terms of both turnover and profitability, and in terms of vehicle quality and reliability.

Toyota was by far the car company with the highest market capitalization at the end of 2018, around $200 billion.[1]

What is the secret of Toyota's success? The incredible consistency of its performance is a direct result of operational excellence, transformed into a strategic weapon. This excellence is partly based on the quality improvement tools and methods made famous by Toyota in the world of industrial production, such as *just-in-time*, *kaizen*, *one-piece flow*, *jidoka*, and *heijunka*. These techniques have contributed to the *Lean Manufacturing* revolution. But tools and techniques are not the secret weapon to transform a company: **Toyota's uninterrupted success in implementing**

DOI: 10.4324/9781003474852-2

these tools stems from a more nuanced business philosophy that is based on an understanding of human psychology and motivation. Its success is ultimately based on its ability to cultivate leadership, teams, and culture; to develop strategies; to build relationships with suppliers; and to promote organizational learning.

In-Depth Analysis

The whole world began to understand during the 1980s that there was something "special" about Japanese quality and efficiency: Japanese cars lasted longer than American cars and required much less maintenance. And in the 1990s it became clear that there was something even more special about Toyota than other Japanese car manufacturers. The difference was in the way Toyota conceived and produced cars, with an incredible consistency of process and product. Toyota designed vehicles faster, with more reliability, but at competitive costs; even considering the relatively high salaries of Japanese employees. Toyota is considered the reference model and the best company in its class by all its peers and competitors around the world, due to its high quality, high productivity, production speed and flexibility. If you want to discover the Toyota Way to achieve these goals, I recommend the book *The Toyota Way. 14 Management Principles from the World Greatest Manufacturer.*

(McGraw-Hill, 2004)

What Can We Learn from the Toyota Production System Method Today?

The Toyota model has been studied all over the world. I too have been deeply influenced by it.

The first lesson I learned was about the Lean Thinker's real obsession with the customer: he is the sole protagonist of any business. The Lean Thinker only cares about realizing, with as few frills as possible, what the customer wants, in the shortest possible time and in the quality that they (and not who produces and sells) desire.

So Lean Thinking is first a different mindset. An attitude that leads straight to the heart of things, to promptly answer a fundamental question in any business: who is our "real" customer? It can be an external customer or an internal customer, our colleague, or a leader to whom we supply a product or service. Whether external or internal, in Lean Thinking we ask ourselves incessantly not only what exactly the value is that customer expects from us, our products, our services, but how can we continuously increase it, while at the same time reducing activities that, on the contrary, do not add value.

The second lesson can be traced back to the unique way that Taichi Ohno, considered the founder of the Toyota Production System, had developed to analyze processes and activities in search of waste: the concept of "Ohno circle." A real circle drawn on the floor inside the factory where he spent hours standing to observe what was happening in the production department. He watched and observed all possible forms of waste.

Unnecessary movements, unnecessary transport, waste, redundant processing, productions not covered by orders, overstocking, waits, and so on. His aim was to identify the causes of waste in vehicles' production, to reduce the time that passed between the moment the order is received and the moment the product is shipped, which would then become the moment when payment is received. There are seven "capital wastes" that, according to Taichi Onno, can be encountered (and cut down) in any production context. The eighth waste was "gifted" to us by Jeffrey Liker in the best seller *Toyota Way*, a book that I had the honor of reviewing and updating for the special 2014 Italian edition: **non-utilization of the full human potential at our disposal**.

The starting point for any radical improvement, in the Lean spirit, is to see our company and our own individual activities through the eyes of the customer. Learn to recognize value in our processes, distinguishing them from the various forms of waste that are possible. Starting to wear "Lean glasses" means completely changing our view on how things are done, being able to be "honestly critical" about everything we do. Every time we set out on a path to value analysis, this can have the power to take us very, very far.

In the Lean perspective, any activity can be of three types.

■ **Value**

This is an activity that the customer recognizes and is willing to pay for or an activity that adds tangible value to the company (Masaaky Yutani added that for him an activity is only of value when the customer is happy to pay for it).

Password in this case: **increase**!

■ **Necessary waste**
An activity that does not add value to the customer, but that cannot be avoided. An example is the procedures for complying with mandatory regulatory and legislative requirements or other activities that may be fundamental in the company despite being "auxiliary," such as the long and detailed activities of drawing up annual budgets.
Password in this case: **reduce**!

■ **Waste**
A completely unnecessary activity, pure waste, for which the customer would never pay and that would make the very desire to buy question itself. The worst thing for this type of activity is that it is often disguised as a necessary or even value-adding activity.
Password in this case: **eliminate**!

Ohno's focus was, therefore, to reduce all waste to maximize company's cash flow. Minimizing the time between obtaining the order and the delivery of the finished product to obtain new liquidity to be reinvested: this basic concept is still perfectly adaptable today in all contexts. Because, at the end of the day, it is a question that prevails over all others: how can we reduce the time taken by materials, information, and resources in our company? In Toyota, they had realized since World War II that the more they could reduce this time, the more money they would collect, thus having a positive impact on the company's accounts. To do this, they were forced to eliminate as much waste as possible, which they saw as real time killers.

If we focused on how we can reduce the time of crossing in different areas, both corporate and private, we would soon realize to focus on reducing various wastes, otherwise not even noticed. Waste that undermines the efficiency of a company's production: new product development, logistics, and administrative.

Fascinating, don't you think? After all, so far, I have shown you the "easiest" part of the problem, the tip of the iceberg. All around the world, flocks of consultants and managers could be able to give the illusory feeling of improvement and waste reduction, as if to retrace the footsteps of the good Taiichi Ohno. And there are now many companies that have done tactical actions of "spot" improvements and optimizations inspired by the good principles of Lean Manufacturing.

Things are not as easy as they seem if you want to get important and lasting results. Definitely not, they are not. Even if the tools described seem so damn sensible and basically simple. In the next section, we will try to understand why Lean Thinking is not easy to implement today. Not even years ago, but today with the strange times we are living in, the story has really become more complicated.

The Dark Side of Lean Thinking

When we talk about Lean and the expectations of the great benefits we can derive from it, there is always a dark side that eludes most people who try their hand at specific programs. **Implementing Lean Thinking in the company force us to face the great change management challenge necessary to achieve consistent and lasting results.** An in-depth study by the prestigious consultancy firm McKinsey, which lasted more than three years, examined and interviewed more than 311,000 employees involved in business transformation projects, including 6,800 CEOs and senior executives, for a total of more than 400 companies analyzed. Lean Transformation projects stood out among the corporate transformation projects, along with organizational re-design, new product launches, and new IT system implementations. The conclusions of the major study were rather embarrassing: in 70% of the cases examined, the projects failed to achieve their initial target and, at most, they achieved good results in the short term but regressed to the starting levels in the medium to long term. The causes of these embarrassing results were also examined: in 39% of the cases, the fateful cause of resistance to change emerged, followed closely, 33% of the cases, by another ubiquitous cause, that of the lack of support for change by managers. Closing the ranking of causes were inadequate resources and available budget, in 14% of cases, and a generic item "other obstacle" in the remaining 14%.

It seems clear to me that we have a problem that is not easy to deal with. And all in all, this evidence should not come as much of a surprise to us. I think that any one of us who has embarked on the arduous task of making a change in our lives has experienced the difficulties that McKinsey has encountered in the corporate world. Maybe we managed to go on a diet at the beginning, not without difficulties, but then after a few weeks or at most a few months, we returned to the starting point. In fact, it is no coincidence that the number of diets that fail far exceeds the number of diets that succeed over time! We may succeed at the beginning with a few more running or gym sessions, feeling proud of having succeeded, but then we inexorably return to our old habits. We could extend the examples to many other fronts of life, both personal and business. Different kinds of projects that start out have some results, but then degrade quickly back to the starting situation, if not worse.

Cursed Resistance to Change

What doesn't convince me is the identification of the main root cause of this performance decay exposed in the McKinsey study: resistance to change. Every time I hear these little magic words I get hives. But what do they really mean? We have all been biologically programmed to resist change. Back in the day, when we were hunter-gatherer nomads, 200,000 years ago, change was perceived as a risk. Any form of new experience brought with it the potential for the discovery of new sources of resources but along with pitfalls from which one must defend oneself: new enemies hidden somewhere, fierce unknown animals, or a lack of food and water. Any human has always had to find a compromise between their innate need for security and their need for variety and discovery. One thing is certain: the more I have established certain boundaries, acquired securities of any kind, whether right or wrong, habits that have solidified, the more difficult it will be to change. This is normal, in life as in business. It is therefore normal for people to resist change! Whatever it may be. And in my opinion the second cause cited by the McKinsey study, the lack of support for change from managers, is not so strong either. It is not a question of support being lacking on the part of managers. The real problem is that very often we don't know how to manage change, we don't know how to find that fragile balance between the need for security and the need for variety, we don't know how to really engage and motivate people, and we don't know how to embrace change ourselves and how to become an example of it.

New Needs to Be Met

The needs of people working in any company have changed profoundly in recent decades. In the past, they correspond to "I work, you pay me," and then maybe I make my "real" life outside the company walls. A basic need for security that is certainly still alive, but that today accompanies others: expression, sustainability, security, connection, uniqueness, comfort, fulfilment, and meaning. These are just a few examples of needs that push people to behave in seemingly inexplicable ways but logically interpretable in the light of slightly more in-depth behavior analysis. Today, people live in a dimension where uncertainty, variability, complexity, the surplus of information, and the speed with which everything happens pose non-trivial challenges, and the more leaders can understand what their people's needs and requirements are, the more unexplored avenues open to create new forms of very fruitful exchange between companies and individuals. And it is precisely these new forms of dual exchange that bring up the performance curve of people and companies upwards. Excellence becomes a daily challenge that is played in a double game: company and person. And if people do not learn to desire excellence, there is no company in the world that can become excellent in its processes, in its products, and in its results. And there is no real excellence if you cannot balance technical and social excellence daily.

You must imagine the company as a tree that grows by sinking its roots on difficult soil in adverse weather conditions. Who wants to collect fruits from this tree, and not just for a season, but for many seasons, you need to take good care of the roots, learning how to protect themselves against adverse weather conditions. The roots are people, the fruits are the results and the value we generate for the market and for all stakeholders. Only with strong, well-tended roots planted in the soil will the tree grow luxuriantly and bear the desired fruits.

Entropy and Energy in the Company

There is another obstacle to the sustainability of the performance of individuals in the company over time that causes the almost inevitable decline in many areas of business activities, including improvement projects that are often launched. Let's try for a few moments to liken our business system, made up of processes, people, and a complex thermodynamic system, that is a system where internal transformations are possible thanks to the interactions and exchanges of matter and energy with the external environment. A thermal engine, two bodies with different temperatures in contact, and an explosive mixture are all examples of thermodynamic systems. Well, every system of this kind in nature suffers from the problem of entropy, an irreversible phenomenon explained in the second principle of thermodynamics. This principle states that when there is no thermal equilibrium, the entropy of a system tends to increase over time until a new equilibrium is reached. This explains why thermal energy (heat) always flows from a warmer body to a colder one and never in the opposite direction. This energy is redistributed until the system consisting of the two bodies reaches a new complete equilibrium, the one in which both have the same temperature, and it is no longer possible to pass heat from one to the other. **With each transformation of the system, the entropy increases, and energy dissipates, unless other energy is added from the outside.** So, if we compare our business system to a thermodynamic system, we can state that each attempt to transform it corresponds to an increase in entropy and thus energy dissipation, which will tend to regress the system performance (Figure 2.1). The moral of the story, if I do not supply my system, and well, with new and powerful sources of energy, it will be normal to

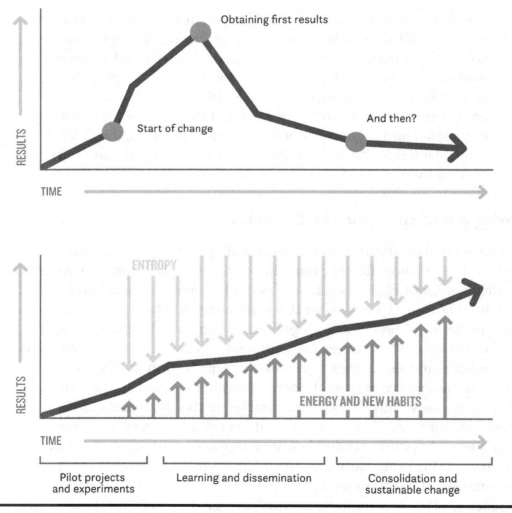

Figure 2.1 Energy and new habits to combat entropy.

witness decay. Any project that alters pre-existing forms of equilibrium brings with it an increase in the system's entropy and thus also the almost mathematical certainty of witnessing a decay in performance if I do not bring in new forms of *energy*.

What are these forms of energy? It ranges from the basic forms of physical energy to the most hidden forms of emotional and mental energy, to arrive at the deepest ones that affect the "spiritual" sphere of meanings and deep reasons why people act.

Yet Does Lean Thinking Still Work Today?

It seems clear to me that talking about and trying to apply Lean Thinking in business today cannot be a simple "copy and paste" exercise. We live in a completely different context than those in which Lean Thinking was born. And the social and economic conditions around companies have been further shaken by the Covid-19 pandemic. Today, the market conditions change much faster than in the past, and demand conditions and customer needs are not comparable with those of past decades and will change even more rapidly in the coming years. Technology offers many opportunities and many risks at the same time for all companies, with a general reduction of the acquisition costs of the last evolutions of the moment business models are changing radically and many new start-ups are literally disrupting some old paradigms, making a lot of money in a different way than in the past. And above all, people are really under pressure, with much more information to manage, huge unpredictable variability, and profound stress conditions to deal with *every day*.

Does Lean Thinking still work? No. Not in the same way as in the past. We need to rewrite some of the rules of the game. We need to dilute the inevitable dark sides that emerge today when we want to apply principles that were born many years ago and in extremely different contexts. We need to start with the current conditions of people and the new needs that individuals and companies have. We must do all this if we want to put Lean Thinking not at the center of philosophical expectations, but at the service of our corporate evolution in the information age and digital transformation. Lean Thinking must be put at the service of people to achieve more results and more well-being in the company and in life. From Lean Thinking to Lean Lifestyle. Together we will see how.

Testimony

"In recent decades we have witnessed the spread of fierce individualism and a culture that has taken competition to extremes. I have been fortunate to work in companies where the application of the Lean culture was authentically interpreted and this represented a curb on the axiom that being the best, even at the expense of others, is a just cause. But after the crisis of 2008, it became more difficult everywhere to maintain the spirit behind the methodology born in Toyota, and Lean has turned a betrayed revolution: the corporate culture based exclusively on the result at all costs, has conquered more and more space, creating a progressive and unstoppable dissatisfaction in the world of work. Gradually, work reverted to be predominantly an answer to basic needs, while at the same time the satisfaction of more evolved needs (such as belonging, esteem by others and the realization of a life that had a higher purpose) was sought in the social context outside work. But now the first global pandemic has significantly reduced or completely cancelled sociality. So how will the absence of emotional compensation outside the workplace impact on people's already precarious motivation? The only possible solution will be to promote a model of leadership that is immune to the god "I" and that aims at the well-being of the organization, because what is good for people will be even better for the business of tomorrow. I believe from my own experience that the Lean Lifestyle approach can be a powerful vaccine to achieve this immunity".

—**Cesare Ceraso,** Former General Manager of Stanley Black & Decker

Note

1. In the summer of 2020, Elon Musk's Tesla Motors surprisingly outperformed Toyota in terms of capitalization, strong of the share price's +500% leap in less than a year. However, the fundamental difference, as of today, is in the numbers: 367,000 vehicles sold by Tesla in 2019 against Toyota's 10.74 million, for real profits of the two companies that are at a ratio of 3.4 to 100, in favor of Toyota.

Chapter 3

High-Impact Work

Always interconnected, but with what consequences? In the age we are living, it has become imperative to be always connected and always ready to respond to all the stimuli and requests received. In this way, however, we run two risks:

- become unaware of the direction that our professional and personal life is taking.
- no longer knowing *why* we do what we do every day.

Imagine for a few moments projecting yourself onto one of the many days you have experienced over the last few months. One in which you felt the strain of working a lot. Not that in the others you didn't, but on that day, you did a bit of the hit. Did you rewind the "tape"? Well, now imagine that you are on the way back or at night, already comfortably at home collapsed on your sofa.

Let me ask you four questions:

Is your effort and fatigue commensurate with your results on that day?

Have you finished the projects you had been waiting for a long time, or have you made significant progress in the desired direction?

Can you tell me if and how your many activities have had a significant impact on your business or personal life?

Can you distinguish and give me a shred of measure about how many hours you have devoted to activities with a high impact for you and your business and how many hours, on the other hand, do you think you have consumed in activities that are necessary, but in the end not really with a high added value, if not even classifiable as pure waste?

DOI: 10.4324/9781003474852-3

I don't know if you could answer all the questions right. Almost certainly the last question will have thrown you off balance and you may not have found a way to quantify the answer with a number. Yes, we all always feel we have done something good, but sometimes we struggle to identify exactly what and especially what kind of impact we have had. We find it even more difficult to say exactly how much time we specifically allocated to the various tasks we took care of during the day. In addition, we are instinctively inclined to say that we have devoted virtually all our time to do urgent, unavoidable, and indispensable things, defending ourselves well in advance against any accusatory acts that would cast doubt on our ability to better manage the time available.

Death Concerns Past Time, Not the Future

The ultimate purpose for which we do all our activities at work, and outside, is to produce outputs in favor of our company, in favor of ourselves or of some other person dear to us, or in response to deep motivations very important to us. However, the way we predominantly work, which is now also becoming the way we conduct most of our lives, does not enable us in a position to understand to what extent we are producing these outputs. Indeed, **we are losing sight of the ability to commensurate activities performed and impacts achieved.** We spend, without realizing it, hours and hours looking at e-mails of relative importance and browsing the web, getting lost in sites that we did not even want to open intentionally. We spend hours and hours in discussions and meetings that we did not want to have, only to be reduced to making very important decisions in a few minutes, even postponing them until the next day, because we are exhausted, complaining punctually that we had little time available to do the "real" work.

We spend our time without fully realizing that every minute spent on an activity has gone away forever.

Time is given to us in abundance because a 24-hour day equals 1440 minutes, which, from my point of view, is a lot of time, if lived in full. And this time knows no social or corporate hierarchies. We all have the same amount every day. However, the hour and minute counter may give us the illusion of repetition and redundancy, this is the illusion that if I reset the 24 hours I had at my disposal, they will be available to us again tomorrow, perhaps to do things I failed to do today. Unfortunately, it is just a mirage.

Time consumed never comes back. Everyone thinks of death as a future event that we try to avert and postpone as far as possible. I think, instead, that death is something we experience every day. Yesterday's time has inexorably gone, along with the part of us that is no longer there.

The Luciano who lived for over 50 years before today has disappeared. He is already dead. Today, I have only inherited from him the impacts and consequences of his actions. They are the inheritance that will influence the way I live my future days. The activities I perform go away and no longer return along with the time I dedicated to them, but the consequences remain. For better or for worse. I may have positive impacts or negative impacts. I don't believe in equilibrium, balance, and neutrality. If you are not producing positive impacts and consequences for you and your business, you are worsening your overall state. Actions that bring positive consequences are not necessarily large and striking, but they must be objectively detectable and quantifiable by anyone, however small they may be.

Every Activity Has Its Price and Every Investment Its Return

Our time is the only real precious and (very) limited resource we have. I know, I am not saying anything new. The point is that the great achievements of humankind did not happen because of what was known, but because of what was done with what was known. How do we use what we have learnt for ourselves and our business? We must think that every time we use the resource of time, we are investing in what is most invaluable to us. We can simply spend it, or we can invest it in activities that have a definite return for us, for our loved ones, for our business, and for our company.

When I touch on this issue in companies, I realize that I am putting my finger in a sore spot that people find to be huge.

It is as if they had bought a fantastic and expensive automatic machine for the production department and could not know how to answer the question about its actual use. Or it is as if when asked about the using in hours of a machine, they answered: 8, 10, or 12 hours, but without knowing exactly how many good parts it has produced.

Yes, it gave outputs, but we cannot say exactly how many or even how many times the machine stopped or slowed down its production cycle.

In good substance, the company would not be able to express itself in what in managerial terms is called O.E.E.—*Overall Equipment Effectiveness*. Consequently, it would not know whether the investment made with the purchase of the machine is having a return or not.

Unthinkable, isn't it?

In all companies, we strive to monitor the efficiency and effectiveness of plants and machines (even if there are sometimes situations that come close to the embarrassing perplexities of this example), but in the face of the enormous human resources available, there is very little to be said about the utilization of the existing potential and the real efficiency with which it is being worked. So much for corporate neo-gurus and all the ready-to-wear consultants who carry and spread the myth of the "person at the center."

If Everything Is Important, Nothing Is Important

> *It is not enough to be happy to have an excellent life. The secret is to be happy doing things that test our abilities, make us grow and lead us to express our full potential.*
>
> **—Mihaly Csikszentmihalyi**

Where can we start to solve the problem, I have outlined? From the clear identification of what I have called Gold Activities, that is those activities that add more value to ourselves and to our business. You will recognize the similarity with the key principle of Lean Thinking: always starting by recognizing the value for the customer in the flow of the activities performed. The difficulty, when we pass from business processes to individual people, is that we must learn to avoid all the pitfalls that prevent us from recognizing what brings the most value, resisting the distractions of various and overcoming numerous obstacles along the way. And human beings are not machines to be regulated and do not move according to workflows that can be predetermined by default.

In the company, even with some difficulties, it is possible to define operating processes, revise them, improve them. And you can, with as many difficulties, bring changes in the behavior of the people who interact in the value streams linked to those processes. With individuals, however, we will have to enter the subtle mechanisms of individual functioning. We will have to learn to distinguish the *thresholds of intentionality*—what I consciously want to do—from those of habitual *automatisms*—what I usually do without realizing it. We will have to stop to understand the behaviors enabled by our functional beliefs versus those disabled by opposing beliefs. If, for example, I am convinced that interruptions get me more results at work, I will pursue this behavior until it becomes automatic for me to interrupt continuously without realizing it anymore. **We will have to make the effort to better understand how we "work" and how our employees "work**." The Gold Activity principle is a "domino" principle because if we fail in our deep understanding of this principle and all its consequences, we will fail in understanding and implementing all the others we will encounter on the Lean Lifestyle path.

The first element to be aware of is a real "superpower" in our equipment and that today is weakening more and more. It is called *focus*.

Our mind works like a camera. Only what is observed exists. The rest does not. Our brain is constantly exploring reality by literally slicing it up and highlighting for our fragile attention span only a few elements at a time. If you are not observing something, it does not exist, even if true. Biologically, we are programmed to protect ourselves from the hundreds of visual, auditory, and sensory stimuli that constantly demand our attention at any given moment.

From an evolutionary point of view, the human being has benefited more than any other animal species from cognitive processes linked to attention. In fact, thanks to the ability to select certain environmental stimuli while ignoring others, attention has been an extremely effective mechanism for survival, organizing information from the ever-changing external environment and regulating reactions and behavior accordingly.

There are numerous studies about human attention, the physiological, mental, and emotional mechanisms that reduce it and on those that can increase it. Different types of attention have been identified: selective, focused, divided, joint, sustained and perhaps more will be identified. Despite the differences of ideas, one element is traversal: our attention span is limited. Daniel Kahneman states that there is a biological limit to the simultaneous processing of information and when this limit is exceeded, a sensory filter is automatically activated that blocks all information that exceeds the processing capacity. In this situation, we lose sight of a lot of information because our mind is only selecting a small number of external stimuli. Therefore, some events we will pay attention to consciously, while others will be registered only unconsciously by our mind.

Therefore, if we are allocating our mental space to certain sources of information and stimuli from outside, it is very likely that there is no more room for anything else, at least on a conscious level.

These considerations lead us toward a deep reflection. During our days, many times we focus on many different things, most of which come from sources outside of us, which literally "steal" our attention, making us lose awareness of the fact that we no longer focus our attention on things that are more important for us.

> Our focus capacity is potentially very strong, but very limited, and it is not at all obvious that everything we focus on is the most important thing for us, our company, and our loved ones.

Focus is like a muscle; if we don't train it constantly, it gets very weak. To be trained, the focus muscle needs a key ingredient: *intentionality*. So, in our case it would be more correct to speak of *intentional focus*. And since most of the sources of focus are currently external sources (e-mails, colleagues, chats, social networks, meetings, interruptions) in reality, intentional focus is already weakened in most business contexts. We live more and more in a rampant state of hyper-reactivity to external stimuli. On the contrary, we become more and more able to react quickly to these external stimuli, at the same time weakening our ability to intentionally concentrate. This leads to a loss of awareness of the impact-activity relationship: if, for example, we are involved almost simultaneously in four activities with completely different impacts on our business, our poor brain cannot easily distinguish the differences in importance. A poor e-mail account will occupy the same time, attention, and energy as a project with a possible millionaire impact for our company. Everything, therefore, becomes of equal importance, or rather nothing becomes substantially more important than something else. In many cases this flattening process does not allow us to observe and choose high value-added activities that we could, and in some cases should, do and instead do not do. We face real *blindness equivalent*. We miss valuable opportunities for ourselves and for others.

The flattening of relevance levels is ultimately one of the behavioral biases encountered in business today. This easily leads to a loss of value, time, and energy. Every day.

Be Focused on Excellence: Gold Activities

The path to excellence starts with the deliberate choice of activities that can lead me more quickly toward value creation. These are precisely the so-called gold activities. But we will not be able to correctly identify any Gold Activity unless we clarify well what excellence and value mean for each of us.

Personal excellence does not coincide at all with the attainment of a standard, pre-established level of performance, as high as one like, but represents everything that leads us to the creation of a value higher than a starting state no matter what it is. Value creation is therefore an entirely individual concept, linked to the individual person who is moving toward his or her personal excellence, or a group of people who are creating greater value than an initial situation to which the group was bound. The word "value" can therefore have several meanings. For example:

- We feel effective and satisfied when we create value for ourselves and others.
- We feel gratified and confident when we perceive value within ourselves.
- We truly feel ourselves when we embody a value.

Therefore, I will find my excellence when I will be able to express my VALUE in full, or that of my group, and this will happen when:

1. **I will have created and achieved something that is important to me**, or contributed to achieve something that is important to me;
2. **I will be able to perceive that I have put to good use one of my talents**, one of my areas of expertise, feeling gratification and personal satisfaction for having done so;
3. **I will feel I have activities in line with my deeper values.**

At this point, a question comes to mind that never hurts to ask again and again:

> Does everything you do really value-added for you, your loved ones, and your business?

A *Gold Activity* is, in essence, a fast track to value creation. One can do many things every day, but it is only through these activities that one will travel the fastest path to personal excellence effectively. Each of us should always know—or should find out as quickly as possible—what our Gold Activities are to achieve the best of ourselves or our group.

You Don't Become Excellent by Doing Everything

You don't grow up by reading as many e-mails as one can or by zeroing your e-mail inbox, as we may have been taught in some time management course. We must distinguish service activities from value-added activities, necessary activities from those that represent waste for whoever does them, and, above all, we must identify the activities that we do that represent an absolute waste.

Gold Activities also allow us to have a strategic vision and not to end up with "our head in the sand" dealing with tactical things every day. A person who concentrates on pursuing his or her own Gold Activities will succeed in integrating the team more and more because he or she will have to choose what to do personally and what to delegate to others, thus forcing the integration needed to manage and monitor the delegated activities. The pursuit of the essential naturally leads to the elimination of the superfluous, simultaneously reducing stress and waste.

Being focused on the activities that we perform excellently and that lead us toward excellence enables us to achieve extraordinary results. When we don't, we create high levels of stress and slowly drag ourselves into a state of exhaustion.

> If we spend too much time focusing on low-value activities, including paying too much attention to our weaknesses, we will end up with a string of strong weaknesses, many activities performed indiscriminately, but which will not allow us to gain a competitive edge in the world of work, making us mediocre at many things, rather than excellent at a few.

Golden Pearls in a Sea of White Pearls: The Gold Activities Scouting

At this point, if the reasons are sufficiently exhaustive, we try to understand how to recognize and accurately identify one's Gold Activities. **Here is a self-coaching path in 8 key questions to ask yourself in the suggested sequence**: the Gold Activities Scouting. I suggest doing this path at least two or three times a year, carving out a private space with high concentration and far from distractions. Under these conditions, an average of one hour or so is enough. You could combine this work with the ATRED work that we will see in Chapter 10.

1. **Where do I want to go and what do I want to achieve in the next 6–12 months?**
   There is always a hidden but indissoluble link between what we achieve in life and the activities we do. The first step in identifying

Gold Activity is to make us "lift our heads" and look straight toward our desired goal. We cannot talk about Gold Activity by remaining solely and exclusively concerned with the tasks to be completed today, tomorrow, or at most by the next week.

2. Are the actions I am doing most of my time getting me where I want to be?

What I am doing now, how much does it help me to get it? An activity is GOLD if it helps me to become excellent! I recommend making a list of your main activities, whether done recurrently or occasionally. Look back in the last quarter or semester at the latest. List everything that comes to your mind and put together the minor activities that seem small to you, but could represent large portions of time: travel, meetings grouped by type, e-mails, authorization signatures etc. We often overlook the fact that several activities may seem negligible in the instant we perform them but become absolutely relevant in the cumulative count.

You will find yourself in front of a generally long list of activities carried out, both work-related and personal. This is the starting point for critical analysis. Some of these activities will have the power to favor the achievement of desired results, others exactly the opposite. And most probably some necessary activities will be almost or completely absent.

3. What should I do more or different than today to get what I want?

After completing the list of prevailing activities you will discover that some activities are already Gold, but you probably do not devote enough time or energy to them. And in most cases, you will find that you do not routinely perform some activities that would be necessary to achieve the results you have defined as the answer to the first question. You can add as much as you like. Don't be afraid, just let yourself go! You are, at this point, in the eye of the storm: you have a huge list of activities, how do you choose the ones that are more Gold than others?

4. Which activities have the greatest impact?

With this question begins the most critical part of the journey and where we must be as honest as possible with ourselves. We must judge the impact of each activity for us, for our group and for our company. We can measure the impact both in tangible terms (e.g. money, sales volumes, measurable performance) and in intangible terms (satisfaction, gratification, professional growth, recognition). You will find that many activities are necessary, but with a small or at least questionable impact, while others are definitely linked to a personal or corporate result of absolute relevance. Other activities may probably come to mind when you go looking for impact. Go ahead and add them to the list.

5. **In which activities am I most proficient?**
   This question makes us take a dip inside ourselves and puts us in front of who we are and our limits, but above all before our strengths. You must go hunting for the activities in which we certainly feel you are putting your best resources, your knowledge, your talents, the skills accumulated over the years. Recognizing the activities in which we are most skilled can open a world of opportunities in the direction of achieving greater results with less effort and more fluidity. Again, if you realize that there are activities in which you are highly skilled that do not appear on your dashboard, add.

6. **In which activities do I have more passion?**
   Free space to search for what makes you vibrate the most, activities that you like or that you would love to do. Discovering what makes you feel positive emotions and that you would never get tired of doing. Here you will discover that many activities you do not because you like them, but because you have to do. It is very likely that at this stage you will decide to lengthen the list with some seemingly unimportant activities, but that have the gift of returning joy and good humor, mandatory ingredient to achieve both performance and well-being.

7. **What activities are challenges for me today? It is about**
   Identifying the activities that definitely take you outside the comfort zone. They might scare you because they will project you into an evolution of our identity, of your current role. But they are the ones that make you the person you want to become. And remember that any company or business sector evolves to the extent that it can evolve who guides and leads that company or that portion of the company. **The answers to this question will make you clearly discover not your weaknesses in general, but those that can limit our development toward the desired goals. It's your *bottleneck*.**

8. **Which of all the activities I have identified are Gold?**
   I must answer this question by counting and measuring the degrees of impact, skill, passion, and challenge of the various activities listed in order to highlight a solid link with the tangible and measurable development of your person and your company. It will emerge clearly that many activities belong to the category of "service," as necessary but not at all Gold. And many activities will also emerge that can be classified as waste, which is to be eliminated sooner or later from the current focus dashboard.

But Do We Really Need All the Gold Activities I Have Identified?

In order to know which Gold Activity "serves" us more than others, it is wise to rely on the old and always valid law of Vilfredo Pareto, an Italian engineer and economist, who stated in the late 1800s that in all complex systems, 20% of the causes cause 80% of the consequences. This principle, also known as Law 80/20, although of a qualitative and empirical nature, finds countless demonstrations in many fields, from economics to social sciences. It all started in 1897 when the engineer, studying the distribution of income, demonstrated that in each region only a few individuals owned most of the wealth. Looking at different areas of your business and life: you will easily discover that from a few customers derives most of your turnover. You could equally easily realize that a few root causes carry most of the quality problems you have. You may discover that a few people occupy most of your time. You may even find that from very few decisions you make, most of the implications of your entire life are derived. If these considerations are true, it will be equally true that **we have absolutely no need to fill our time with so many activities to make the greatest impact in our quest for results and well-being or redeemable so many gold activities.** We need to improve the selection of a few activities that will make the difference. Our golden pearls are in a sea of common and white pearls.

It's a change of perspective on how we look at work. A focus on how important the personal aspect is in professional life and vice versa.
Serena Apicella, HR Director Peroni

Useful Feedback That Comes from Afar

Another valuable element to consider in order to be sure that we are on the right path toward the adoption of Gold Activities concerns our own perception of time involved in doing various things. The ancient Greeks distinguished time into two broad categories: *Kairòs*, God's time, that during which something truly special happens; *Kronòs*, chronometric time, the during which ordinary things happen, not at all special and objectively monitorable with external devices or supports. Back then it might have been an hourglass or something similar, today it's a watch or a chronometer. In *Kairòs* mode, we perceive that time differently. Remember how you felt when you were immersed in an activity you enjoyed and were so passionate about that you even lost track of time. A very exciting hobby, a fantastic walk in a forest never seen before, a long kiss with a loved one, a particularly beautiful and engaging job. In these moments time seems to stop or run very slowly, and, above all, we forget to have a watch. In *Kronòs* mode, time must constantly be observed and controlled. An uninspiring meeting, an activity that you just don't like or that at least doesn't excite you. In all these cases, we feel the pressure on us of the flowing external time.

> A real gold activity should make you fill in full Kairòs mode. If you struggle to intentionally enter this mode, it means that your work system does not guarantee you spaces of depth and concentration without distractions and interference.

In the following chapters, we will see in detail how to create a working system that can lead us down this path.

> Traditional tools are no longer enough. We must have new eyes and tools to read the complexity. A path of improvement that starts by themselves and then must involve the company.
> **Paola Artioli,** CEO Asonext S.p.A.

Gold Activities Never Shout

Let us try to draw conclusions from this chapter. By now it should be clear that **the problem to be solved is not summed up with the question "how can I do so many things in the same time frame?" but by "how**

do I learn to do fewer things in the time available?" The "right" and most impactful things, of course. There are people who have created empires and reached the peak of their specialties not doing more than others but by acting differently from others.

> It is important to disengage from the common conception that the result depends on the volume of activities carried out, just as it is vital also to be wary of the over-commitment to which we are accustomed.

When you hear the most popular phrase of all "I don't have time, I'm too busy," there is something wrong. Often when people say they have no time, all they are doing is hiding the real problem: I struggle to choose the right things to do and I find it even more difficult to get rid of so many things that I find myself doing today, without making it perfectly and even when I don't want to or shouldn't. Time, I repeat, is the most democratic thing there is, it is the same for everyone, and the real difference is not who manages it better than others, but who uses it differently from others.

Gold Activity leads us on a path of healthy real productivity, which diverges from the path of frenzied commitment. It is always better to aim to be very productive rather than very busy.

We are all over-committed, but being productive and being able to choose the things that increase our value and that of our group and our company is another story!

Remember that a Gold Activity does not impose itself as urgent; it waits for you to choose it, and that is why it differs from "background noise" or from all other activities that by their nature will always ask about us (e.g. phone calls, other colleagues, e-mails, text messages). We intentionally choose and make it. Very often we give up because it is more convenient to do the same thing as always, complaining about it, rather than stopping and deciding to do something we have chosen to do. One should not fool oneself by declaring oneself too busy to do something more profound and high impact. When we do that, we give up on our most important challenges. It takes courage to take the road of change. But it is our choice that will trigger the excellence around us and in our company.

The chapter that is about to be concluded had the objective to focus on the importance of choosing key activities to achieve our excellence. In the next chapters, we will see how to concretely free up space in our lives and how to eliminate constraints and waste that prevent us from finding time and energy to dive into our Gold Activities.

Lean Lifestyle Story: Gualapack

Starting from the Awareness

The great absence in the daily agenda of Italian managers: time to think. And not only. Time for employee development, time for priority planning, time for training, and, even more serious, time for strategy are also at the margins.

The evidence of this widespread criticality also emerged during the initial stages of a Lean Lifestyle growth project for 30 top managers of Gualapack, the Alessandria-based company specialized in flexible packaging and injection molding solutions, present with its 10 plants in 7 countries (Italy, Romania, Ukraine, Mexico, Costa Rica, Chile, and Brazil).

They were asked to list the three main gold activities they regularly did and the three main gold activities they did not do enough.

The reclassification of the answers revealed that the most popular Gold Activities indicated by almost all managers, beyond 80%, were strategy, development of team potential, and priority planning. However, the first two were not performed at all, or only rarely, by about 50% of the managers.

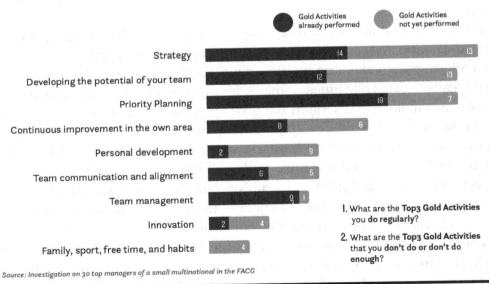

ARE WE DOING THE RIGHT THINGS?

Source: Investigation on 30 top managers of a small multinational in the FACG

Figure 3.1 The main Gold Activities of Gualapack managers, classified between those performed regularly and those not performed enough.

Almost all of them therefore shared what should have been their priority activities, but half of them were equally aware of not doing enough to play their role to the full.

Lean Lifestyle Story: Ferretti Group

The Impossible Challenge of Working Almost 100 Hours a Week

Reflecting on Gold Activity is never an obvious exercise. Sometimes you even get to question the very way you understand your work. This is what happened to **Alberto Arfilli**, plant manager for the Ferretti Group, Italian pride in the design, construction, and sale of luxury motor and pleasure yachts, with a portfolio of eight prestigious and exclusive brands: Ferretti Yachts, Riva, Pershing, Itama, Mochi Craft, CRN, Custom Line, and Wally.

I have always been a person who liked to change. I have held various roles in the company, with increasing levels of responsibility. Until October 2020 I managed the Forlì and Cattolica plants, then I have been called to the guide of the Ancona plant and today I have the responsibility of five yards and almost 600 people. When I started working on my ATRED I first structured a project of delegation and growth of my staff starting from an extraneously "burdened" personal situation: I worked almost 100 hours a week, Monday to Friday, a couple of hours in the morning from 5 to 7, then 12 hours in the plant with a half-hour lunch break talking about work with colleagues. Then a queue in the evening from 10 p.m. to midnight and at the weekend I continued working, usually waking up early in the morning, but with no set hours, for about 7 to 8 hours, both on Saturdays and Sundays. The family and personal sphere obviously suffered. The rules I had given me allowed me to manage two production plants, but they had become expensive and heavy to bear, even physically. I needed to acquire tools to ease the burden, otherwise the alternative would have been to "give up" something.

The first real step in this direction, however, were not immediately the tools (which came later): to be decisive was the reflection on what his Gold Activities really were. Initially, he had identified 19 Gold Activities that he considered fundamental. By better analyzing his role, he realized that he no longer wanted to be the kind of plant manager he considers all those important activities. He realized that some of the operational activities that he considered Gold at first were not even appropriate for him, while some aspects related to the growth of employees were completely absent from his list.

At the end of this second reflection, he chose only nine Gold Activities for himself, some of which were not even on the initial list. And on 3 of the starting 19 he ended up delegating them to his first reporters, thus deeply questioning the very way he interpreted his role. Moreover, this process of analysis also meant reflecting on personal Gold Activities and the impact those work rhythms were having on his life and health.

> After less than six months, I got home at 6 p.m.; no more working at home in the evenings and at weekends. Opening work e-mails has become an exception.

How did he do it in practice? Let's find out through his story in which we will anticipate tools and ways of working that you will find detailed in the continuation of the book. Now simply assessing the extraordinary impact on professional results and quality of life.

> Once I understood the Gold Activities, the delegation was the natural consequence. I started with moderation. Some activities were completely delegated to my first reports and together we also redesigned and standardized some of the reporting and related activities. The site visits have also completely changed in structure; the total time has paradoxically been reduced, but they are now much more effective as we do them every day with a specific focus in rotation, and I try to make them a time for confrontation and growth rather than a time for "control." I then reviewed all the meetings I ran to make them more effective, and I cadenced all the meetings so that everyone's agendas had organization and clear references. For example, the time for my weekly staff meeting was reduced by 50% thanks to a standard shared agenda with a defined and respected duration, initial preparation, and a different organization of the dynamics of development. Moreover, thanks to the new habit of preparing the report containing the actions to be done, I have eliminated the onerous task of having to draft it up after the meeting (which used to take me up to 2 hours). And by

working on the empowerment and sharing the objectives with others, I also eliminated the time-consuming activity of checking and reminding them of the actions that I used to do in the following days. Paradoxically, both were things that I first thought were my Gold Activities!

Alberto then introduced "Sacred Time" into his agenda, dividing it into two phases: the first has a "private" form and takes place early in the morning, a time when planning the working day, and the second at the end of the working day, a time when taking stock of the workday.

The Lean e-mail management was another major achievement: the core of the improvement task was to remove all notifications, read them only at certain, desired times of the day, and reduce the flow of e-mails exchanged thanks to the changes and delegations introduced. All this led to a reduction of more than 50% in the time spent (from 15 hours per week to 7).

> By reducing all this waste, I was able not only to practically halve my working time, but also to increase the effectiveness of my action and, above all, also to be able to give more time to a Gold Activity that I was not doing, that is to invest significant time to the development of my collaborators and to regular one-to-ones with them: I did not do this before and now I dedicate at least an hour a day to it.

What happened?

The quality of my work has improved and so indirectly the company's results have benefited. I freed up time and energy towards activities of value for me and the company. Thanks to the path made with Lenovys, I was able to grow from a managerial point of view. In previous plants I left behind more rewarded and responsible travel companions. And I've always been careful that they can also benefit from adopting more agile ways of working, otherwise I wouldn't have established a virtuous circle in which everyone could achieve results and well-being.

Some people have grown and occupied new roles of responsibility, also thanks to the delegations they have been able to use. Others I brought with me in this new experience, making them grow from the point of view of work. These are the greatest

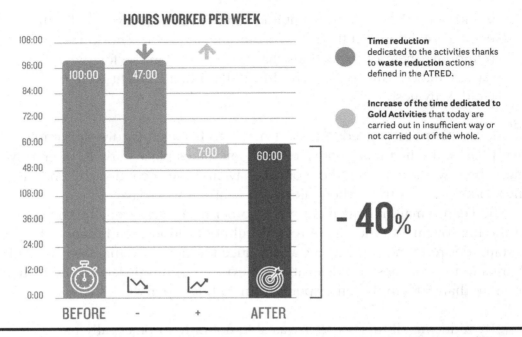

Figure 3.2 Summary of the results of the improvement project by Alberto Arfilli, Ferretti Group plant manager.

satisfactions, seeing the people who work with you grow. And me? I continue to apply the Lean Lifestyle methods, with satisfaction, and put them into practice even today, that my production reality has changed. The responsibilities remain, but I have lightened the workload. I have learnt that you don't need to work 14 hours a day to achieve results and have no problems in your business, and that a different organization of work can make us achieve more and better, for ourselves, our family and the company.

Lean Lifestyle Story: Chromology

Starting from Yourself to Change an Entire Company

One of the turning points for my growth as a leader and for the development of the company was the Lean Lifestyle Master's course I attended in 2016. The evolution that has arisen on an individual level and on the level of my employees, many of whom followed the course in the following years, and the consequent change of pace on a strategic level, helped to build the Cromology of the future. A company totally transformed because, despite the difficulties encountered along the way, we chose to act with a clear strategic and no longer tactical approach.

Massimiliano Bianchi, since 2015 head of Cromology Italia (part of Nippon Paint International Group), leader in the production and sale of building paints (his brands include MaxMeyer, Duco Baldini Vernici, and Settef), has changed the way he works to change the way the company does business.

I started from myself, from the way I manage my priorities, with a goal: to dedicate my valuable time to customer care. I wanted to meet them personally with the ambition of giving a human face to our multinational and thus becoming their partner. To do this, I worked on defining my Gold Activities, managing my agenda, and delegating. All of this made my way of working more focused and already in 2018 allowed me to organize the first big meeting with 700 customers, in which we communicated to all of them the new strategic path based on the Partnership concept, and today I can proudly say that I personally go to meet many dozens of partner entrepreneurs every year, whereas before I didn't get to two.

Time to make strategy. This is the most important Gold Activity for Maximilian:

I needed to find mental space. Freed from activities that were not relevant for my role and goals, strategic thoughts and energy for focused action began to flow in my head. Among my new habits, I can no longer give up the half-day blocks I schedule weekly for reflection and analysis. Naturally, we have extended the Lean Lifestyle training to my co-workers and started internal courses to disseminate the working methods learned. Among the most obvious changes were the management of meetings: from an unstructured approach, we have moved to a Lean organization and conduct of meetings, in which preparation (assignment of tasks, choice of participants), reference standards (agenda, reports) and focus (no use of smartphones) have increased the value for people and the company. Another structural action taken was the scheduling of management, departmental and individual one-to-one alignment activities. A year in advance, the entire organization has recurring appointments on its agenda, and this has simplified and made the activity of preparing the meetings less stressful.

These and other structured initiatives to improve the energy and well-being of people in the company, such as providing managers with an in-house coach for individual support, were recognized in 2018 with the award of the Great Place to Work prize.

Well-being and results: from 2016 to today Cromology has consolidated its market leadership and tripled its EBITDA percentage. And all thanks to a new way of working and doing business where agility, simplification, and people's quality of life have come together to improve individual, team, and company results.

Lean Lifestyle Story: Lucchini RS

Gold Activities as "Medicine" for Mental Well-Being

Davide Raineri is the head of the Process Technology and Estimating Office in the Railway division at Lucchini RS, a company operating in the steel and railway sector, the largest supplier of wheels, axles, and wheelsets for the global high-speed train market.

As part of an extensive Lean Lifestyle® training program in the company has launched a personal improvement project that led him to change perspective on his way of working.

> The keystone in my approach to the Lean Lifestyle® has been the Gold Activities. When I started to understand the meaning, I had an epiphany: "Damn, I'm not doing them!" I knew those activities were important, but numbers in hand the time I was devoting to them was a big zero. I said to myself: if I have time I'll do it, otherwise "peace." But thinking about the concept of Gold Activity, a light bulb went on. Apart from the time, because they were activities for a few hours a week, paradoxically, when a project or an unexpected activity came along, one way or another I completed it, whereas for those I could never find the time. Making a list of the activities I do and quantifying them helped me to focus on the problem. In particular, I found a new habit very useful

that has quickly become a Gold Activity: the weekly Friday evening hansei that I dedicate on the one hand to sorting out the next week's planning and on the other hand to think about what I can improve. It really helps my "sanity" and makes me enter the weekend more serene. Obviously, all this was made possible and sustainable by waste reduction actions (for me mainly related to e-mails, interruptions, and standardization), which allowed me to save more than 2 hours a day that I could convert into professional and private Gold Activities.

Chapter 4

What Makes Work Really Difficult

"It is different here. In our industry, in this company, you cannot change . . . You cannot understand." How many times have you said or heard these statements? Unfortunately, identifying the few high-impact activities is only the beginning of a new way of working. You must understand why even the best intentions can flounder in the sea of today's working world.

The Basic Problems

Are you sure that your operating strategies guarantee you maximum output in the shortest possible time and with the minimum expenditure of energy?

Are you sure you know, and above all know how to make the best use of, the techniques to achieve maximum efficiency in the processing and production of your intellectual content?

These are two questions that we should get used to answering several times in our lives, because as the context around us changes along with our evolution, the answer does not necessarily remain the same over time.

DOI: 10.4324/9781003474852-4

The Three Deceptions of Our Mind

We must, however, be very careful. To make the situation more complicated, there are some basic problems valid for everyone: which are cognitive biases that mislead us about the answer we might give to the two questions above. Let's see them.

1. **The cognitive-behavioral deformation: zoom in vs. zoom out**
   Human beings inexorably tend to overestimate what they can and must do in the short term and underestimate what can be done and achieved in the long term. It happens when we give in to a temptation that gives us a small instant gratification, rather than waiting for a large time-shifting gratification. When we are assailed by the anxiety of dealing with all the topics that come to our mind and for fear of losing some of them on the way and delaying their execution, we address them all immediately. With the result of completing very few of them and being perpetually late. Or, even worse, when we enter a state of *action paralysis* that blocks us and makes us procrastinate endlessly. We no longer realize that we are dazzled by the lights that appear to us in the immediate moment and that prevent us from seeing beyond, with the effect of hardly ever being "here and now," truly present and with the lucid perspective of those who observe the entire time horizon in front of them. Our brain cannot function well with several windows open at the same time and easily succumbs to immediate gratification if we do not train to have temporal perspective, that is the ability to **focus well on what we want to achieve in the next hour and what we want and can achieve in two, ten, a hundred or a thousand hours of work**.

2. **The cognitive-behavioral deformation: subjectivity vs. objectivity**
   We are not able to be objectively aware of the micro-quantities of time lost every day that add up to the huge amounts of time. Our subjective perception is almost always wrong. Think about how many times you give in to your mobile phone and take an innocent little "trip" rummaging through WhatsApp messages, e-mails, and various social media. That time seems little to you, and it almost feels like a well-deserved relaxation break: what eludes us is the total cumulative time we ultimately devote to these innocent little escapes. According to research conducted by Mediakix, a leading US marketing agency, in 2017 every American spent

about 1 hour and 56 minutes per day on social networks.[1] The data were collected from a segment of the common population, excluding professionals who, for professional necessity, must make use of these communication tools. What does this time correspond to? If we do our math right, it corresponds to almost 1 in every 10 years of life. Do we really have nothing better to do than throw it on social media? The point, however, is that our brain cannot quantify the entire time it will be throwing away. Our instantaneous subjective perception will almost never coincide with the objective quantification of the total sum of time spent in the long run. And this does not only apply to the social networking issue but to all **the little things—of which we are not fully aware—that apparently represent a trifle, but objectively represent huge, accumulated amounts of time**: television, web, phone calls, distractions, interruptions, useless reading, useless relationships, useless activities, and so on

3. **Cognitive-behavioral deformation: rhythm vs. chaos**
   The third basic problem relates to the absolute underestimation of the rhythm with which things happen in our lives. The human being lives by biological rhythmic elements, for example: heartbeat, sleep-wake rhythm, menstrual cycles, and hormone production cycles (melatonin, cortisol, testosterone, etc.). Whenever we comply with these biological needs, our body gives thanks and creates an alliance with the brain to achieve maximum effectiveness with minimum energy expenditure. Turn your attention for a moment to what happens daily during your working days. Do they seem to you to be the realm of rhythm and paced rhythms or a rollercoaster with sudden braking and violent acceleration without any warning? Often at work we are inundated with innumerable inputs that come to our attention without warning, without order, without any rhythm, and at the same time the sequence of required outputs does not follow any criteria: we pass with indifference from moments in which we have to give 10 answers at the same time and do not know how to do it, to moments of apparent calm—a few— in which we feel almost lost and wonder if there is something wrong. The point is that it has now become normal to accept these ways of working, not realizing at all the negative effects that there are first on individual and group intellectual performance, as well as on the state of psycho-physical health. **Working in a constant state of stress and hyperreactivity, in fact, causes a state of biological alteration** that we will explore in more detail in the following pages.

If we really want to put in place behaviors to streamline and simplify our days, we must first understand how to cancel at the root of the three basic problems we have just seen, and which are literally bedeviling the working days of us 21st century intellectual workers.

Mind Like Water

Can you get into a hyper-productive state at will?

Imagine you are sitting on the banks of a mountain pond. Crystal clear, flat water. Silence in the air, perfect quiet. You are enjoying that moment. Now imagine picking up a pebble and throwing it into the water. What would happen? A series of concentric circles would form from the exact point of impact of the pebble with water. The circles would first increase and then gradually disappear, returning the pond to its pre-throwing quiet condition. This phenomenon is well known to martial arts experts. The concept of mind as water refers to the ability to keep the mind calm and clear to understand when it is time to act with extreme precision and effectiveness, striking the right blow at the right time and then return to the state before the blow.

Just as water reacts neither disproportionately nor less than required, the power of a karate shot is the result of rapid execution and its precision in impact, not muscle strength.

> Don't get set into one form, adapt it and build your own, and let it grow, be like water. Empty your mind, be formless, shapeless like water. Now you put water in a cup, it becomes the cup; You put water into a bottle it becomes the bottle: You put it in a teapot it becomes the teapot. Now water can flow, or it can crash. Be water, my friend.
>
> Bruce Lee

In order to exercise this control over mind, there must be no wandering thoughts, distractions, and loss of focus but a state of presence and strong concentration, "here and now," which allows one to carefully read the opponent's moves knowing how to foresee their actions. Free mind and flexibility, these are the secrets of martial arts.

Well, after this digression, let us return to our offices and our daily work.

How many times has it seemed to you that your mind is like water? How many times have you been able to enter this state of mind at will?

Our mind is like water, but not exactly that of the good karateka. It is a turbulent water, always agitated, always moving from one subject to another: it is never still and motionless because it is struck by dozens of pebbles at the same time. And the movements on the surface are not drawn as concentric circles but by confused intersecting shapes.

When the water is agitated, it is difficult to see clearly. When it is calm, on the other hand, everything is clearer.

If we want to see better and come closer to the precision of action of the karateka in the world of intellectual work, we must understand how to regain control of our mind. In the following pages, we will analyze both the most common causes of mental agitation and the main solutions to bring it back to a state of calm.

At the Root of the Problem: Multitasking, Joy, and Pain in Business

If you chase two rabbits, you will not catch either one.

Russian proverb

During the master's courses I conduct, I do experiments on quantifying the loss of time we incur when we perform several tasks at the same time and when we change topics frequently. I have found the same results in thousands of people: a significant amount of time is silently wasted due to the continuous change of focus and tasks to be performed. Why does this happen? Why do we lose these huge amounts of time when we change the task or topic we are working on? And above all, why do we not realize this, paradoxically thinking that we are more productive in the so-called *multitasking* mode?

Imagine being "immersed" in an activity called "A" (Figure 4.1) when, for whatever reason, you are forced to shift your attention to a different activity, which we call "B." This apparent change of state costs us time and mental energy, which is called "transition time" or "switching time." This transition time represents a real fee, which does not wait for your consent to be paid.

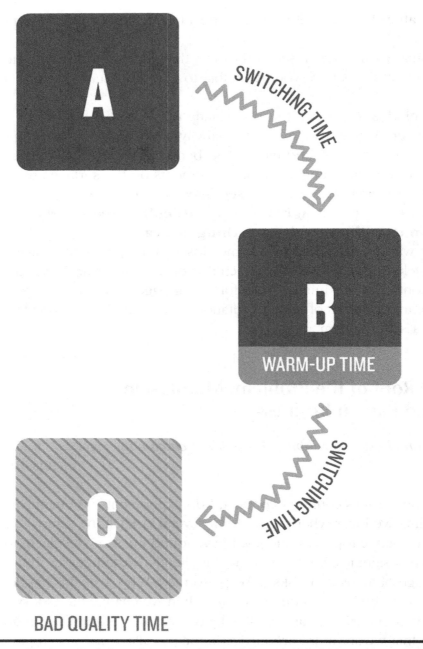

Figure 4.1 The "cost" of changing attention and focus: switching time, warm-up time, and bad quality time.

Transition time is a price we pay for every change. Always. However, its duration may vary depending on the type of activity that has been interrupted.

In addition to this lost time, there is also another type of destroyed time: the so-called "warm-up time," that is the time needed to fully concentrate on the new activity B. If A and B were two electrical appliances to be connected to the electricity grid—our nervous system—the warm-up time would be the time needed to disconnect one from the network to attach the other: I can do this as quickly as possible, but the time taken will never be zero.

According to the studies by Earl Miller, neuroscientist at MIT, one of the world's leading experts on human learning, attention, and cognitive abilities,

> when we switch from one activity to another, the process often seems fluid and energetically free, but in reality, it requires a series of small shifts in focus. Each small shift involves a cognitive cost. For example, every time you move from handling e-mails to writing an important document, you are exhausting valuable brain resources and energy.[2]

A designer, for example, when he goes from designing an "A" product to designing a "B" product can take 15 to 23 minutes to regain the same cognitive efficiency as before the interruption.[3] Our brain is comparable to the household appliance mentioned in the previous example. But it is a single-function household appliance. That is, it can only perform one operation at a time. If I am doing A, I cannot do B at the same time: I have to stop A, pay the time and energy duty, and then start B. Moving from the metaphor of the household appliance to that of the computer, our brain corresponds to the only CPU in our possession, the central processing unit wired into a microprocessor, which superintends the main functions of the computer. Why do I speak not only of lost time, but always of lost energy in the task change? Because my nervous system has to perform some neuronal readjustment and orientation operations to prepare itself for the new task: it has to connect the neurons prepared for the new task, activating the right synapses needed to perform the new functions, and it has to activate the appropriate memory areas to provide us with all the information we need to perform the new task. Whatever it may be. To us, these all seem like immediate and zero-energy operations but try to answer the following questions.

How do you feel after changing your attention and focus on so many different topics and tasks?

Do you feel tired or fatigued compared to when you did only one activity or a little more in the same period?

To complete the picture is a third category of time destroyed daily, what I have called "bad quality time." What is it? Think about how many times in your day you do something, but with little concentration, or think about several things at once, but without paying attention to anything, or when a person talks to you, but you have hardly listened at all. These are a few examples of how we have become accustomed to filling our time with many small chunks that drain mental energy, but add no value to us, nor to those around us, or our company. PS: the *bad quality time* that makes me smile the most is when we answer the phone to simply tell the caller that we cannot talk because we are busy with something else. But why not automate this message and have it gone off autonomously whenever we have decided not to use the phone? It would save us many unnecessary interruptions.

Human Brain and the Ability to Achieve

In 2017, Adam Gazzaley and Larry Rosen, neuroscientists, professors of neurology and psychiatry at the University of California, published an interesting book summarizing much of their research on the functioning of the nervous system in highly distracted contexts. Our nervous system at about 200,000 years from its last great leap forward, the one that led us toward the birth of *Homo sapiens*, has remained essentially the same, with its basic functions unchanged. We are constantly looking for positive or negative signals from our surroundings in order to process the information needed to make the decisions that will guide our actions. The fine regulation of this sophisticated mechanism, briefly described in Figure 4.2, has allowed us and our loved ones to feed, procreate, and avoid being killed in a prehistoric world full of pitfalls. The major difference between our brains and the more primitive brains of all other animal species lies in their sophisticated decision-making capacity. Other animals have no brain processes that enable them to make deliberate decisions and no capacity for formulating high-level goals that can guide their behavior. They are guided entirely by their reflexes: external environmental elements activate sensory neurons through specialized receptors that send signals to other neurons responsible for predetermined motor responses.[4]

As we also saw in the previous chapter about focus and our attention span, there are limits to our ability to process large amounts of data and information. The "superpower" that Mother Nature has

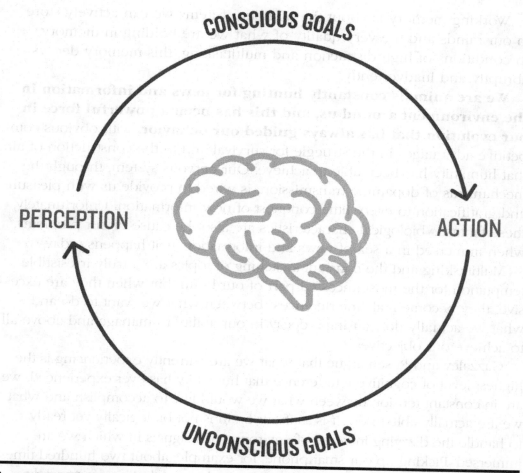

Figure 4.2 Diagram of the perception–action cycle in the human nervous system.

endowed us with seems to be physiologically limited due to three physical "bottlenecks":

■ attention;
■ working memory;
■ ability to manage different goals.

The limits of our attention capacity manifest themselves in the fact that we are forced to select everything we want or need to deal with from time to time and in the fact that distributed forms of attention lead to much lower cognitive performance than the focused type of attention. The sustainability of attention over time is limited and the speed of processing changes heavily depending on how we are using our attention "tank."

Working memory is about the number of items we can actively store in our minds and the very quality of what we are holding in memory. In conditions of high distraction and multitasking, this memory decays abruptly and inadvertently.

We are animals constantly hunting for news and information in the environment around us, and this has been a powerful force in our evolution that has always guided our behavior, with obvious competitive advantages in the struggle for survival and in the construction of all that humanity has been able to achieve. Our nervous system, through the mechanisms of dopamine transmission, is wired to provide us with pleasure and gratification to every little conquest of new information. Unfortunately, these fantastic biological characteristics are also the cause of our problems, when immersed in a sea of news and information as it happens today.

Multitasking and the constant switching of topics are a truly irresistible temptation for the most ancestral part of our brain, but when they are excessive, they become real "interferences" between what we want to do and what we actually do, causing a decay in our ability to manage and above all to achieve our objectives.

Gazzaley and Rosen argue that what we are currently experiencing is the highest level of cognitive interference that humanity has ever experienced: we are in constant tension between what we would like to accomplish and what we are actually able to do. It is as if our brain is not biologically yet ready to handle the dizzying increase of environmental signals in which we are immersed. Picking up our smartphone, for example, about two hundred times a day, on average every four to five minutes when awake, is unfortunately not an act without consequences. We go on tilt and suffer, without being fully aware, a limitation in our capacity for fulfilment. It happens at work, when we want to learn something, in our relationships and in all other daily activities where decisions and actions are at stake mediated by the elaboration of the context around us. That is, in almost all the meanders of our life. This tension cannot fail to have serious mental, physical and emotional consequences.

When Does Multitasking Increase Performance?

Researchers have shown that cognitive performance is not the same in all situations. We can distinguish three cases:

1. When we feel bored or unmotivated by the performance of a particular task, our performance will not be particularly brilliant.

2. There are, on the other hand, situations in which even with the concomitant performance of several tasks, we feel more excited, inclined to action and particularly brilliant, experiencing a type of stress that is functional to our performance.

3. Finally, there are situations, in which we inadvertently slip into an excess of stress that severely limits our performance.[5]

The Yerkes–Dodson law, represented in Figure 4.3, shows that as what the two researchers called "arousal," stress, or excitement, you get an improvement in brain function, up to a point of no return, after which there is a drastic drop concomitant to a state of high anxiety and stress. This point of no return, however, is much closer than we think.

Several researchers have tried to better understand how to recognize the stress peak that triggers deterioration. One among many has piqued my interest. It is a study, published in 2015 in the journal *PLOS ONE* and conducted at the University of Florida. The research in question was conducted on elderly people who had to complete cognitive tasks while cycling on a stationary bike. An improvement in pedaling speed was observed if they performed simple cognitive tasks (for example, pronouncing single words) and decreased as the tasks became more difficult. The researchers attributed this slight improvement to a kind of facilitating effect of multitasking associated with an increase in *arousal* in the person. In this case, the consequence in the brain is a release

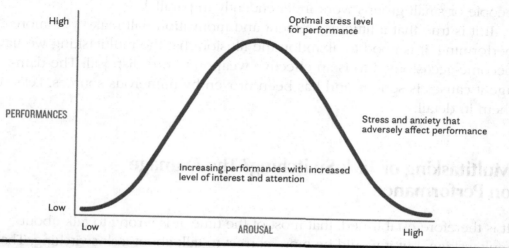

THE YERKES–DODSON LAW
How anxiety and stress affect performance

Optimal stress level for performance

Stress and anxiety that adversely affect performance

Increasing performances with increased level of interest and attention

High

PERFORMANCES

Low

Low

AROUSAL

High

Figure 4.3   The Yerkes–Dodson law.

of dopamine, norepinephrine, and epinephrine that improve speed and efficiency of the brain, particularly in the frontal lobes. These effects increase the availability of additional cognitive resources that facilitate performance in both motor and cognitive tasks with a mutually reinforcing effect. But this phenomenon is particularly relevant in cases where the starting situation is one of impairment of dopamine transmission mechanisms in certain areas of the brain, as in the case of people with Parkinson's disease and in the elderly.[6]

Outside of these circumstances, on the other hand, the starting conditions are markedly different from those with Parkinson's disease. On the contrary, the starting levels of dopaminergic transmitters are already quite high in the stressful working environments in which we are immersed.

The benefit of performance multitasking can only exist when one of the two tasks is something I have already accomplished hundreds of other times, so that I perform it almost automatically, without even thinking about it and without external variables to pay attention to.

For example, if we go for a run on a treadmill or on an exercise bike—and we are used to doing this—we can read or talk on the phone or watch a movie without big snags, in fact this will make the exercise less boring. But as soon as one of the two activities gets a little complicated, the idyll ends.

Therefore, unless I use my brain at very low speed for one of the concurrent activities, the only multitasking that provides serious benefits is that of performing parallel tasks that do not engage brain activity, for example: I turn on the dishwasher and go cook lunch. I will have two parallel activities going on, washing dishes, and preparing lunch. Another example of effective multitasking in business is when I manage to multiply the brains in use for carrying out tasks and projects, for example when I delegate well and let several people or small groups work independently in parallel.

If it is true that a little excitement and motivation will make you more performing, it is good to abandon the illusion that the multitasking we have become accustomed to is an effective weapon to our disposal. The damage it causes is serious and has been proven by numerous sources. Let's see them in detail.

Multitasking or Task-Switching? The Damage on Performance

It is therefore established, that most of the time, it is wrong to talk about multitasking, but it would be more correct to talk about task switching. That is, task switching rather than multitasking. The real issue is not whether we work in parallel or in series, because in practice we almost always work only in series, but how frequent the task switches are and how long the time losses are after each interruption. This is exactly what happens in a production line on which several products must be manufactured. To maximize productivity, one tries to minimize inefficiencies due to product type changes, reducing setup times and rationalizing the frequency of setups when it is difficult to reduce the periods of exchanges. Humans are among those production lines where it is impossible to reduce change durations to zero.

Let us assume a working day of 8 hours and 20 minutes. We will have 500 minutes to manage and put to good use to advance projects and tasks toward their goals. In the following pages we will see that some studies show that the time losses due to frequent changes of subject can reach 50% of the total time available. Let's assume we are much better at our performance from "multitasker" or "task switcher" versus average and consider a loss of only considering a 20% of time each time we change topic. This will mean for example, about 120 seconds every 10 minutes of interrupted activity, sum transition and warm up time to move the focus from task A to task B and back from task B to task A or go to another yet. With these simple assumptions, if we imagine changing tasks every 10 minutes, we will have a dry loss of 100 minutes in a whole day consisting of 50 slots of 10 minutes each. That is just over 8 hours a week, 1 whole day burnt for nothing, 4 days a month, about 50 days a year: about 2 months a year!

If, instead of interrupting our flow of activity every 10 minutes, we tried to stay focused for 30 minutes, for the same duration of unit loss per change, the weekly time-wasting would be reduced to less than 3 hours, the monthly time-wasting to about 1.5 days and the annual time-wasting to about 17 days.

With the slot duration of 60 minutes, instead of 10 minutes, we will have much smaller losses, amounting to 1.7 hours per week, less than 1 day per month and only 10 days per year. Therefore, it is impossible to zero-time losses due to multitasking, but you can reduce them through the right ways of working.

Figure 4.4 shows the three cases in a time slot of just over two hours. As you can see, the percentage of time lost due to topic changes has much less impact on the total as the duration of the single slots increases "mono task."

Unfortunately, **research has shown that 11 is the average number of minutes a worker may devote to a specific project or task before being interrupted**, and that all interruptions have a negative effect on performance, also increasing every time the inaccuracies and errors during the

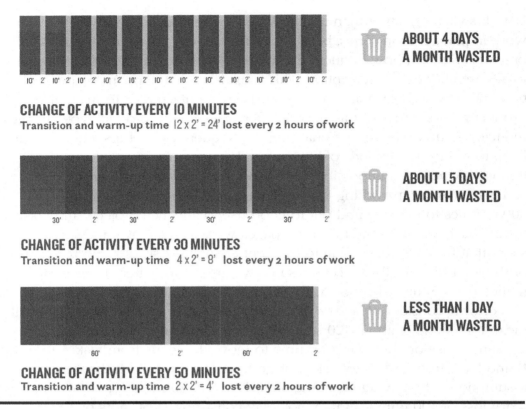

CHANGE OF ACTIVITY EVERY 10 MINUTES
Transition and warm-up time 12 x 2' = 24' lost every 2 hours of work

CHANGE OF ACTIVITY EVERY 30 MINUTES
Transition and warm-up time 4 x 2' = 8' lost every 2 hours of work

CHANGE OF ACTIVITY EVERY 50 MINUTES
Transition and warm-up time 2 x 2' = 4' lost every 2 hours of work

Figure 4.4 Comparison of the effects of task switching between different ways of organizing work.

execution of the most frequently interrupted tasks.[7] The same researchers have shown that in contexts characterized by numerous interruptions, workers have to work much harder to complete the various open tasks, resulting in a significant increase in stress, effort and energy expended, and feelings of pressure and frustration. The US scientist author of the research, Gloria Mark, herself, when asked to comment the consequences on the brain of our hyper-connected lives said:

> When we started this research about ten years ago, we saw that people in the office switched from one screen to another every three minutes. Then we repeated the study again six years ago, and we saw that people were in front of a device, before switching, for 1 minute and 15 seconds. When we did the study again a short while ago, we found that the time in front of a screen collapsed at 40 seconds.[8]

Professor Mark's studies have shown that every time we are working on something important, like writing a report, and suddenly we stop, pick up the phone, and start chatting or checking e-mails, it takes us an average of 25 minutes to regain our concentration: a mode that has a very high cost in terms of stress and money.

According to the US consulting firm Basex Research, these continuous interruptions result in 28 billion man-hours lost by companies in the US alone, about 2.1 hours per day on a 40-hour weekly basis, about $588 billion every year.[9]

In addition to the already mentioned switching and warm-up time, in everyday working reality, we have to deal with a third category of burnt time, which is underestimated by most people: the one we already knew a few pages earlier as "bad quality time." An example: I am in the office, a colleague comes in and asks for my attention, I say yes, and then turn my attention to him, listening to him. Then the phone rings, I answer it asking my colleague to be patient, I absent-mindedly talk to another person on the other end of the phone, I go back to my colleague from before, I listen to him again, but I am always thinking about the proceeding phone call. And in all this, time flows inexorably, but with a very poor level of quality and concentration. I could add dozens of other examples. Think of meetings with a huge level of distraction due to reading messages and news on digital devices or sitting in front of the computer with one's head elsewhere. Smartphones, computers and all other perpetually connected digital devices have transformed our lives. They allow us to work everywhere, shorten distances and handle countless small daily tasks, theoretically freeing up time for more interesting things. The amount of information each of us receives is constantly increasing. And have we, in these last years of evolution, acquired strange superpowers to receive, filter and retain all this information without consequences? The evidence tells us exactly the opposite. The term "information overload" was coined by Bertram Gross in 1964, a professor of political science. This was Gross's definition:

Information overload occurs when the amount of input to a systemexceeds its processing capacity. Decision makers have limited cognitive processing capacity. Consequently, when information overload occurs, it is likely that a reduction in decision quality will occur.

The concept of *information overload* was popularized by the American futurist writer Alvin Toffler in his famous book *Future Shock* published in 1970. Paradoxically, the more we move into the information age, as Alvin Tofler called it, the more our difficulties increase because we are literally bombarded by a huge amount of information that requires our attention minute after minute. And our brain is not prepared for this overloading of information and stimuli, becoming more and more "excited" but increasingly inefficient in completing its tasks. We open many, but really many, doors, but we close fewer and fewer.

A rather cruel synthesis of the damages on the company performances caused by multitasking unleashed by the mismanagement of all inputs received comes from research by Realization Technologies, a US company active in the development of technology platforms for complex project management[10]:

- -40% on our productivity;
- +50% time to complete our open activities;
- + 50% chance of making mistakes;
- $450 billion/year lost by US companies due to multitasking (data 2010–2013).

Technology provides the tools that should make us more productive. The problem is that we mostly use them incorrectly: people are not trained to use them, and managers do not know how to handle who have to use them.

J. B. Spira, CEO of Basex Research

Why Is It Difficult to Get Rid of Multitasking? The Damage on Health

We spend days, months, years jumping from one interruption to another, feeling that it is only external events that drive us, never vice versa.

That multitasking is not an accelerator of our performance, we may have realized. But the hoax, in addition to the damage, comes from the analysis of the consequences for our state of health. Neuroscience research, such as that of Stanford University recently published in the Proceedings of the National Academy of Sciences, is demonstrating how multitasking progressively worsens the performance of the brain, which loses lucidity, memory, and organizational capacity in constantly jumping from one thing to another. Clifford Nass, one of the professors leading the research conducted on a sample of 100 students, says "everything distracts them," referring to the loss of concentration ability that heavy multitaskers suffer and that prevents them from being able to distinguish important information from completely irrelevant information.[11]

But if this blessed multitasking is so harmful, why can we not really be aware of these inefficiencies, convincing ourselves of the opposite? Recalling the cognitive bias "subjectivity vs. objectivity," it is important to observe a phenomenon: in the continuous change of topic our brain, as described above, is over-committed to so many functions. It works hard to meet our demands for continuous change of focus, so much so that we feel very busy, to the point of feeling fatigued after a few hours in this working mode. That is what deception is. We feel so busy and so fatigued that we paradoxically feel more fulfilled by the effort expended than by the results achieved. Having, moreover, inherited beliefs that come from afar, which reward effort more than results, we also end up commiserating and metaphorically patting ourselves on the shoulders in these situations.

We are also not used to measuring our performance at work, monitor output and results. We are more inclined to embark on long and tiring hours, working most of them in sub-optimal ways. Finally, our lack of awareness is greatly increased by the vicious circle in which we fall, fueled by the physiological phenomena that are triggered in a chain in an environment of numerous interruptions.

Interruptions and Dependence

Scientific research shows that constant interruptions due to e-mails, phone calls, and the surplus of incoming information change the way we behave and think. To every new stimulus we receive, the reference is to a primitive impulse to respond at once to immediate opportunities or threats. The stimulus causes arousal—a rush of dopamine—which, according to researchers, can be addictive. According to Daniel Levitin, professor of psychology and behavioral neuroscience at McGill University in Montréal, Canada, multitasking creates a vicious circuit of dopamine dependence, because it provides instant gratification to the brain when it loses concentration and when it is constantly seeking external stimuli. Levitin claims that the same brain regions that act when we want to stay focused on a single topic are easily distracted by external stimuli. So, whenever we end up in multitasking mode, surfing the Internet crazily, scrolling through social media feeds, checking e-mails, etc., we train our brain to lose focus and become distracted. But it doesn't end there. Just like under the effects of a drug, our brain becomes addicted to dopamine rush every time we change the subject and lose focus. And when the addiction takes hold, it is very difficult to break it.[12]

Dopamine is an important neurotransmitter, with functions of control over movement, the so-called working memory, the sensation of pleasure and reward, prolactin production, sleep regulation mechanisms, certain cognitive faculties, and attention span. In a work context predominantly characterized by tension and nervousness, the release of dopamine brings salvific, transient, and immediate feelings of relief, which repeated over time create the addiction explained by Professor Levitin. And this scientifically explains, in part, why it is so difficult to break out of fragmented work habits. I say partly because unfortunately the negative effects do not stop there.

Attack and Escape

In multitasking mode, our brain works in a special way: the receptors of its reptilian zone are stimulated, activating the sympathetic vegetative nervous system, which predisposes us to a completely reactive behavioral mode, In other words it prepares us to react to external stimuli and we even go crazy looking of them because our nervous system is on the alert and in high tension in search of dangers and opportunities. We are then overwhelmed by a river of neurotransmitters such as adrenalin, noradrenalin, and cortisol, the very precursors of the classic "attack-escape" reaction. As our ancestors did 200,000 years ago, these hormones put us in the right conditions to survive and save us in dangerous situations. Well, in these conditions, the functioning of our prefrontal cortex, which presides over the brain areas of logical-analytical reasoning, and which allows us to stop and think about the most effective strategies to implement, is largely inhibited.

Researchers at the University of London have even shown that in these situations our IQ drops by up to 15 points, or the IQ level of an 8-year-old child, as if one had just smoked marijuana or stayed up all night.[13] On the other hand, excessive intelligence is a luxury we cannot afford in a dangerous situation, faced with a fierce beast, slow thoughts are not allowed, but only quick thoughts, recalling the terms coined by Daniel Khaneman, winner of the Noble Prize for Economics in 2002. Attack or flee, access only to memory areas where we can quickly grab archival information useful in an emergency. The longer we stay in multitasking mode, the harder we struggle to get out of it. A terrible vicious circle that profoundly alters our cognitive and physiological state, the persistence of which causes damage now demonstrated by several authoritative sources: tendency to hypertension, hyperglycemia up to pre-diabetic conditions, heart problems, memory damage, deterioration of cognitive, and semantic functions. **People who practice many activities in parallel for a long time have great difficulty concentrating**, are no longer able to select the various sources of external stimuli, and can no longer overlook irrelevant information, to undergo high degrees of psycho-physical stress.

Everything Leaves a Trace

In addition to all this, research is finding that, even once the parallel activities are over, a "fragmented" mode of thinking and difficulty concentration persist. Two neuroscientists, Kep Kee Loh and Ryota Kanai, from the University of Sussex, have demonstrated through a study carried out on 75 people a disturbing physiological effect that explains the above difficulties when using several multimedia devices for a long time in multitasking mode: the density of gray matter in the area of the brain known as the "anterior cingulate cortex" is reduced, as can be seen in Figure 4.5, which depicts MRI images of the brain structure of the subjects being studied.[14] What does all this mean? That the area of the brain thought to be responsible for controlling cognitive and emotional functions deteriorates. And this explains why people subjected to long periods of media device indigestion in multitasking mode suffer negative impacts on memory, cognitive, and

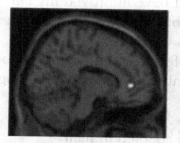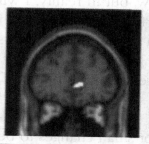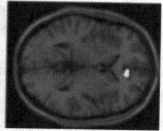

Figure 4.5 **Examples of functional MRI images by neuroscientists Kep Kee Loh and Ryota Kanai.**

semantic abilities to the point of depression and chronic stress symptoms. Other studies have shown that it is possible to achieve exactly the opposite effect or thickening of gray matter density in the anterior cingulate cortex through specific mental training. One example of this is the case of London taxi drivers in the pre-satellite age, who were forced to remember all the intricate streets of the city by heart to do their job. Another example is given by meditation, which has been associated with a beneficial effect on our brain exactly opposite to that of multitasking.

> They told me, but I never would think it was such a powerful gesture to trivial: remove notifications from my mobile phone! Being able to focus without distractions on one topic at a time has drastically reduced the time necessary to exhaust it but, above all, allowed my interlocutor, often my collaborator, to feel devote the time and attention that deserved. Really, it was priceless.
>
> **Lorenzo Bonacina,** Plant Director Marcegaglia

How Can Interruptions Be Minimized?

The picture that emerges from what has been examined on the effects of multitasking seems bleak: we work really hard but with poor efficiency due to constant interruptions, and as a reward we take home an embarrassing series of physical and mental problems.

The good news is that in this framework, every small improvement will have significant results on our performance. If our overall performance does not exceed 20–30%, that is if we work 10 hours to complete only 2–3 hours of real added value, a small increase in efficiency will be enough to appreciate the benefits. If I go from 20 to 40%, the overall efficiency will still seem low, but in reality, I will have improved my current productivity by 100%! Imagine what it can bring to produce twice as much output as today with the same total hours worked. It means that I can seriously start thinking about working far fewer hours or think about much more ambitious strategic goals or simply about drastically reducing current stress and fatigue.

> The secret is to look for small changes to be adopted in their working methods without pretending to make big revolutions. Go hunting for actions that make us gain 10–15 minutes at a time, until you total 1–2 hours per day gained and at the same time regain

energy, concentration, and well-being. On the other hand, being able to earn 1 hour a day means earning 5 hours a week, 20 hours a month, 30 working days a year.

What would you do with an extra month and a half of life gained each year? I leave that answer to you. I have thought about mine . . . It's time to act with countermeasures to interruptions at both individual and organizational levels.

Testimony

Fabio Camorani

Electrolux Lean Manager

When I first attended the Executive Master Lean Lifestyle Leadership, I didn't really know what to expect, other than support in the direction of personal excellence I was beginning to undertake. Instead of helping I had tools, inspiration, strength, energy to open the drawer of personal dreams. Although I was already heading in the right direction, thanks to the Lean Lifestyle I have reached personal levels I did not imagine. A tremendous boost.

The dream drawer has become Pandora's box. And everything has changed, in personal but also in professional growth, as a natural consequence.

My pursuit of personal excellence has also gone through a radical change of diet and sporting goals that I considered only a dream. And as a next step, I passed on all that I had learnt to my team. I tried to involve them, finding very fertile ground, and seeing many of them change, following my example. A unique feeling, a very high satisfaction. I don't think I could have felt like a leader-coach without the Lean Lifestyle.

Lean Lifestyle Story: Zhermack

Reduce Interruptions at Work and at Home

The line between private and working life is becoming increasingly thin. The interference in the space and family time of e-mails and tasks that should remain in the office is total—every evening and on weekends.

This is the reason why **Luca De Simone**, Global Sales Manager of Zhermack, part of Dentsply Sirona, an US dental equipment manufacturer and dental consumables producer, has rethought his way of working from a Lean Lifestyle perspective.

"Living a chronic condition of overload," he recounts, "the first step was to apply the technique of slotting and batching to email management, a task that used to take up several hours a day of my time; in parallel, I worked to reduce multitasking and be sure to have identified the 2–3 key priorities in every day."

The actions taken risked colliding with the fact that for the work carried out he had to be available in the evenings and at weekends, at least for emergencies, and this inexorably led him to have his work phone with him and to look at his e-mails even at home.

The solution was simple: I started forwarding my work phone calls to my private phone, from which I have no access to my company emails. In this way, I was able to stop using my work phone at home and on weekends, while remaining reachable by phone for emergencies. In doing so, I stopped seeing emails in my private phone and regained well-being and energy in the hours and days when it is right to recharge my batteries.

As a good commercial, one of the issues to be addressed was that of travel. The first step was to remove certain trips from the agenda that could be avoided and to introduce a schedule that allowed for more "decent" travel, without absurd combinations and schedules. "In the end, paradoxically, being more rested allowed me to be even more effective."

Awareness. My Lean Lifestyle journey began with an individual awareness process, later extended to my group, through in-depth and impactful training on the topics of personal excellence, high impact work, simplification, energy (physical, mental, emotional), people development, habits. Then, guided by a Lenovys mentor, we turned awareness into action by starting to apply Lean Lifestyle principles to challenge the way we work.

I understood and analyzed the ways I worked, multitasking and interruptions first and foremost for years wrongly associated with productivity, efficiency, and availability. First I suffered them. Now I decide what must happen, what can interrupt me, when and how and how often to be available, exclusively and dedicated, for my team. It was difficult for me and my team to leave old habits behind, but the knowledge helped us to build new, more effective ones, and eventually we found a different way of working that we still maintain today. Have I eliminated multitasking altogether? Not completely, I confess, but that's normal. However, today it is infinitely better than before. And above all, when I fall into the trap of multitasking, I am aware of it and I know that there is something wrong, and that makes a difference too.

Today, my agenda is guided by Gold Activities in my weekly calendar: thanks to slotting on Monday, I have the certainty that 80 per cent of the activities that are important to me, and my work will find the right space and only 20 per cent of my time can be filled by unplanned activities. Among my Gold Activities there are the One-to-Ones with co-workers. Before they were often improvised, almost "left to chance" or to urgencies, whereas now they are an activity that has structured moments "in cadence" within the agendas. Even strategy today has fixed slots in the week in "sacred time" mode where I delve deeper, study products, read market reports, think. I have also started to carve out more and more space to increase my skills, to study and delve deeper. Over the past two years I have completed a master's degree at an international business school, a critical certification for my performance as a manager, and I am enrolled in a new master's program to further deepen my knowledge.

I remember that in the early days of my individual Lean Lifestyle project, caught up in the enthusiasm for slotting and Gold Activity, I created a very dense and rigid agenda full of sacred time slots. It lasted a week before I was forced to blow up everything: anyone who wanted to meet me would find the first available slot in a year's time! But I didn't give up, the spirit I learned to adopt was that of the experimenter: I adapted my slotting and other techniques until I understood what I really needed and how to make them work in my case.

Now I made gold even for the sport . . . Over the years I have reduced my running outings. Now it has a privileged place in my slotting: in my diary I have 2 Pilates sessions and 2 tennis sessions per week, I often take long walks after lunch, and I still do run at the weekend alongside my new hobby of cycling. When I exercise I not only recharge physically, but the awareness and concentration applied to the activity I do helps me not to think about anything else and this helps me to recharge mentally as well.

One of the lessons that has allowed me to keep firm the new habits that I have put to fruit is that it is precisely in times of difficulty that you have to hold on to the new Lean Lifestyle habits.

When the work pressure rises, you need to be lucid, to focus on the things that make a difference, to put in place all the strategies that allow you to avoid wasting time, to work better, to be able to fit in moments to recharge your energy (with sport, for example, in my case).

Do I work more today? No, probably less, but it is mainly the way that has changed. I work better because I have sacred time, and this has also spread to my co-workers so that we can all be focused on all times.

In summary: the Lean Lifestyle allowed me to understand that the way I worked at was not optimal for me, my employees and for professional results. After this step, the rigorous method of taking action to adopt new ways of working that guarantee more results and greater well-being allowed me to achieve the first positive results. With the results I gained, I had the satisfaction of seeing the change bear fruit and this triggered further changes and even more awareness. And eventually this becomes your way of working.

I was addicted to interruptions and now I am in control of them, I can work better, with higher quality and with full satisfaction in my work and private life.

Ermanno Delogu, CEO in KUKA Roboter Italy

Notes

1. Andrii Sedniev, *Insane Productivity for Lazy People*, 2019, www.andrii-sedniev. com
2. E. Miller, Here's Why You Shouldn't Multitask, *Fortune*, 2016, https://fortune. com/2016/12/07/why-you-shouldnt-multitask/
3. Gloria Mark (Department of Informatics University of California, Irvine, CA, USA), D. Gudith, U. Klocke (Institute of Psychology Humboldt University Berlin, Germany), *The Cost of Interrupted Work: More Speed and Stress.*

4. Adam Gazzaley and Larry Rosen, *The Distracted Mind: Ancient Brains in a High-Tech World.* MIT Press, 2017.

5. Francesca Gino, Are You Too Stressed to Be Productive? Or Not Stressed Enough? *HBR*, 2016.

6. L. Altmann, E. Stegemöller, A. Hazamy, J. Wilson, M. Okun, N. McFarland, A. Shukla and C. Hass, Unexpected Dual Task Benefits on Cycling in Parkinson Disease and Healthy Adults: A Neuro-Behavioral Model, *PLoS One*, 2015.

7. Gloria Mark (Department of Informatics University of California, Irvine, CA, USA), D. Gudith, U. Klocke (Institute of Psychology Humboldt University Berlin, Germany), *The Cost of Interrupted Work: More Speed and Stress.*

8. Interview given during the Rai 3 program, 'Presa diretta' on 15 October 2018.

9. Jo Averill-Snell, *Is Multi-Tasking Counterproductive?* American Management Association, January 2019. www.amanet.org/articles/is-multi-tasking-counterproductive/

10. Realization Study: Organizational multitasking costs global businesses $450 Billion each year. www.prnewswire.com/news-releases/study-organizational-multitasking-costs-global-businesses-450-billion-each-year-221154011.html—PRNEWSWIRE 2013.

11. A. Gorlick, Media Multitaskers Pay Mental Price, Stanford Study, *Stanford News*, 2009, https://news.stanford.edu/news/2009/august24/multitask-research-study-082409.html

12. Daniel Levitin, *The Organized Mind: Thinking Straight in the Age of Information Overload.* EP Dutton, 2015.

13. Janssen et al., Integrating Knowledge of Multitasking and Interruptions Across Different Perspectives and Research Methods. https://discovery.ucl.ac.uk/id/eprint/1465496/

14. Kep Kee Loh and Ryota Kanai, High Media Multi-Tasking Is Associated with Smaller Gray-Matter Density in the Anterior Cingulate Cortex, *PLoS One*, 24 September 2014.

Chapter 5

Simplify Working Days and Save Precious Life Hours

The experience I've had with myself, my company, and with hundreds of companies I have worked with leads me to state that there are no magic recipes for getting rid of constant interruptions, for simplifying and fluidifying chaotic working days and activity flows for individuals and entire organizations. There are, however, dozens of small stratagems and operational techniques that can make a considerable contribution in this direction and bring tangible benefits. The techniques I describe below have found multiple levels of application, from the individual to the entire organization. They certainly represent key elements in setting a new way of working.

1440 Mindset

Our behaviors derive from the reality that our mind has modeled through beliefs, thoughts, projections, and interpretative filters on everything we have experienced and are going to experience. This model represents the connecting bridge between us and what lies outside of us. In every moment of our life, our brain carries out real exploratory missions in the outside world in search of elements to confirm the model that has been created. Our emotions, what we feel on a visceral level, often without the apparent possibility of control, are closely linked to the contradictions and experiences we have during these exploratory missions. An example. If I am deeply convinced that I do not have enough time to do everything I should, I will act accordingly

DOI: 10.4324/9781003474852-5

and desperately try to do, in the little time I think I have available, as many things as possible so as not to leave any behind. The prevailing emotion I will experience with this representation of the world I am immersed in will be of anxiety, a sense of urgency, and a sense of hurry. In this situation, by adopting non-functional work techniques, it will be very likely that I will not be able to finish all that I would like to, and many things will be left behind, confirming the initial conviction that I do not have enough time to finish all my tasks. The sense of anxiety, urgency, and hurry will continue to be fed, into a perfect vicious circle. Let's try to break the paradigm.

What would happen if you were deeply convinced that you have plenty of time to do everything?

And what if the many things I feel I must do are actually very few?

The answer to both questions is: I would behave differently.

My mind would calm down and probably become as still as the water I mentioned at the beginning of the chapter. I would behave like someone who knows he has all the time he needs to do the things he has decided to do, crumbling the previous sense of anxiety, rush, and urgency. I would not overlap the different things on each other but give each its necessary time. Let's think about it: every day we have 24 hours of 60 minutes each, which is the beauty of 1440 minutes. Have you ever tried to be out of breath for 1 or 2 minutes? Is it a few or is it a lot? Have you ever tried meditating for 5 minutes? Is it a few or is it a lot? Have you ever tried to stay focused on a single topic without any distractions for 10–15 minutes? Is it a few or is it a lot?

The 1440 mindset is a prerequisite to all the other countermeasures we will see, because it sets us up with the right mindset before diving into action. It is often said that the amount of time available is the most demo-cratic there is because it is equally distributed for all people living on this Earth. We all have 24 hours or 1440 minutes per day available. True. However, it is not the quantity of time that makes the difference in our lives, but the quality of this time. We have a great opportunity to live 1440 pre-cious minutes every day. Let us set ourselves the goal of living them fully and intensively, burning as few of them as possible.

Selective Silence

We live continuously submerged by background noises that never cease to attract our attention and stimulate our mind and our reactions. From the noises of the rooms, we live in to the noises of the various multimedia devices that now accompany almost every moment of our lives. All the noises coming from outside are a reason for loss of attention, loss of concentration and are a perennial cause of interruption compared to whatever one is doing.

As if they were not enough to take us away from here and now the many internal noises, that is, all the distractions and defocusing thoughts that arise continuously inside our minds.

So, one countermeasure that we would do well to put into practice right away is to train ourselves to be selectively silent. That is, to impose upon ourselves periods of silence "on command." Just as the "1440 mindset" represents a cognitive prerequisite that prepares and predisposes us to the most functional behavior, selective silence represents an indispensable condition to work with maximum effectiveness. Start by removing the sound signals from your smartphone, tablet, and PC, or alternate moments with the devices' silent mode with moments with them ready to interrupt you at will, without having any control over them. I remind you that even a phone call you receive that you decide not to answer is an interruption. If you do not practice selective silence, perhaps starting with small portions of time that then become larger and larger, you will never be able to stop your mind "on command." This decision of yours might create turmoil in someone who on the other side of the phone no longer receives immediate hearing and response. But it will be good for you and for others because it will be the beginning of a new way of managing the timing of mutual communication. Get used to entering automatic replies such as "sorry I am busy, I will call you back as soon as I can" or "I can't answer now, I am completing an important job, I will do it as soon as I finish." Get used to creating such rules, and get used to educating others in this way too. If you don't start it, no one else will do it for you. If you let your phone ring all the time; let every e-mail that arrives invade your space; and let every WhatsApp message, text message, LinkedIn, Facebook, Instagram, and Twitter notification become your instant priority, it will be very difficult to realize what your real priority is at the moment and you will end up accepting the fact that you are constantly being interrupted, without being able to do anything about it. It's up to you, just you. Nobody is forcing you to keep everything on.

I can already hear voices of protest from those who tell me that it is impossible not to respond immediately when at work and that this would be an unacceptable revolution. I do not believe in "all black or all white." And like any change, it will be difficult to go from all on to all off, suddenly.

You decide what and when to keep it off. You decide it. But train yourself to do it. This choice will give you unexpected benefits in terms of both the effectiveness you will gain and the well-being it will bring to your stressed synapses.

Sacred Time Blocks

Think of a free climber who is climbing a rugged rock face 2000 meters high, without any protection. In your opinion, can he get distracted? Can he wander elsewhere with his mind? Can he interrupt his activity to do something else, perhaps to answer the phone?

Think of a conductor who is conducting a group of musicians in a concert, under the spotlight of a grandiose stage with hundreds of spectators. Do you think he can afford to think about anything else? Or interrupt his activity to do something else, perhaps to answer the phone?

Think of a surgeon who is performing open-heart surgery on a dying patient to save him. Do you think he can leave the instruments in his hand and listen to a colleague who asks him for an opinion on the therapy given to a patient in hospital?

I could continue with examples of doctors, athletes, musicians, artists, writers, designers, and other professionals who are forced to live a temporal dimension absolutely free of distractions and with very high concentration for the effective performance of their activity, to avoid any tragedy or failure of their execution. They live in a condition of "sacred time block." A time in which multimedia devices "fall silent," distractions disappear, thought becomes focused only on the activity being performed, and interruptions are not contemplated, except possibly as a form of small regenerating pause when a drop in concentration is felt that may undermine the outcome of the activity.

Without ifs and buts, this is *how* great human endeavors are carried out.

Can we fully translate this mode of operation into our offices and businesses? The answer is no, but we could benefit enormously if we cut out one or more blocks of sacred time in the day. We could start with a block of sacred time of 20 minutes, schedule it when we are reasonably sure that we have minimized the chances of interruptions from others, and switch to the selective silence mode to focus only on one thing I have chosen to do. One thing only, abolishing any possibility of distraction and interruption. In those 20 minutes you are very likely to produce more output than in a standard working hour, with medium to low concentration and numerous interruptions due to colleagues, phone calls, messages and so on. How long can a block of sacred time last? According to Tony Schwartz, a well-known US business productivity expert, the sacred time block work mode can even lead us to halve the length of our working days, remembering, however, that we can be effective at high concentration on a single topic for a maximum time of about 75 minutes,[1] after which it is always advisable to take a regenerating break of 5–15 minutes. This way of working can be applied both at the individual and team level, with the precaution that it is composed of a small number of people—max. 4/5—focused only on one topic with high concentration.

Hard to think of isolating oneself in the company and working focused without interruptions?

A question of habits and culture. As with any new habit, the imperative is to start with small but steady steps. The example of leaders who begin to self-organize in this way, to protect their blocks of sacred time and to dedicate themselves to their colleagues and collaborators at specific times, scheduled and focused only for them, will be the winning weapon in the company for the acquisition of this new habit at an organizational level. Always being available for everyone is the most deceptive and inefficient mode of operation you can adopt.

> Sacred time as a personal resource to find high concentration and achieve results faster and with greater well-being. At Streparava, after specific training on individual productivity tools, the transition to the construction of a company routine to encourage high concentration work made it possible to root in the behavior of over 70 people the habit of carving out at least half an hour a day free of any interruptions to devote to developing new skills or working on strategic company projects.
>
> The appointment with sacred time was initially set in the 1.30 to 2 p.m. slot for all employees engaged in office work. This corporate choice made it possible to quickly consolidate the new habit, and once it had taken place, employees were given the freedom to block off a slot of half an hour, or more than one, of sacred time on their agenda throughout the day.
>
> When the individual benefit became apparent to all, no one turned back. It became normal to isolate oneself in one's office and work at high concentration. It is also adopted in remote working mode when one wants to isolate oneself for a few hours during the week and chooses to work from home.
>
> **Raffaella Bianchi,** Human Capital Manager Streparava S.p.A.

Avoiding Overloads and Managing Energy During the Day

Imagine a runner about to run an important race or test over a long distance. There are certain mistakes they will try never to make. One of these will concern the distribution of their energy during the course. The runner knows that if they start too strongly, they will struggle to make it to the end and that if they start too slowly, they will have to rely on a final sprint to achieve a good time and position. The stronger and more advanced the athlete becomes, the more they acquire the ability to manage and optimize their energy during competitions and training, dosing their efforts and listening to their body from time to time, because they also know that one day's performance is never the same as another day's. The same athlete will try to eat well to get the right amount of energy and the necessary nutritional elements and will stay away from foods that can weigh them down both in the short term, in the run-up to a race or training, and in the long term, to make their overall lifestyle sustainable. We enter a competition every day that we work. Whether we like it or not, our working environment has become—and will increasingly become—demanding, high-intensity, with demands that put us through long, demanding hours.

It is unthinkable not to acquire the same mindset as the athlete just mentioned and to work without dosing our energy during long daily marathons. But in most cases, I don't pay much attention to both individuals and companies for managing energy and overloads. Have you ever had to face a complex and important problem at the end of the day, while you have solved only urgencies and held less important meetings the rest of the time? Have you ever taken part in meetings where you have had to wait until the last 15 minutes to deal with the most important issue on the agenda in a hurry, without perhaps even finishing it? I'm sure it has happened to you, too, that you've had to do a great deal of e-mailing first thing in the day, topped off with scrolling through your Facebook or Instagram wall, with the effect that you *enter the field* already intellectually overexcited, but a little disoriented and with great difficulty in maintaining high concentration in the following hours.

It will have happened to all of us a few times to feel exhausted as early as mid-morning and delude ourselves that the cause was a "simple drop in sugar" that we remedied with coffee and snacks on duty. The fundamental problem in all these examples is that when we work, we confuse "*feeling busy and tired*" with the actual performance delivered. A mistake that the athlete of before is very careful not to make. They have a merciless stopwatch that measures the results of their efforts. After all, none of us is in these conditions at work. No one is in the condition of the production line operator where the quantities to be produced have been calculated in advance and assigned according to the technological characteristics of the machine and according to the operations they must perform. Far be it from me to want to mechanize and standardize intellectual work, not least because it is impossible, but the goal that interests us all is still to learn how to achieve more with less effort! **Our energy during the working day is not constant**; we are not robots that just need an electrical connection or a power battery and that's it. We are complex human beings of flesh, bones, and a heavy, energy-intensive brain working incessantly for us.

On Awakening

From the moment we wake up, after the hours of sleep in which everything has been silenced, we have different options of using our body and our brain. We will have different effects depending on the choices we make. We can predispose our body and mind to enter a high-performance, high-energy condition or predispose them for fragmented, high-energy modes of work.

The introduction into one's day of a morning routine that promotes maximum performance and well-being is a widespread practice among the top performers of any discipline. This morning routine can include several elements: meditation, breathing exercises, stretching and mobility, writing one's "logbook," exercises focusing on the day's priorities, a healthy and nutritious breakfast, gratitude exercises, etc. The goal in the early morning should be to take the reins of one's day into one's own hands right away and help body and mind get into the best possible shape. In none of these routines appears the use of the smartphones and reading e-mails and social network feeds, because these are actions that disconnect us from ourselves and project us from the moment we wake up into a world of inputs and stimuli over which we have no control, highly stressful and hyper-exciting cerebrally. They are activities that would immediately take us into a reactive, fragmented, and impulsive state of mind.

On the Working Day

Going forward in the working day we should, both individually and as a team, always try to plan the most important and strategic thing as the first task to be performed, placing the less strategic, but necessary elements in more functional time slots.

> In the hours when we are most tired, we should plan work elements with reduced concentration intensity. Or at least a reduced need for decisions to be taken.

In fact, our brain has a real decision-making reservoir available that is emptied daily with every decision we make, no matter how big or small. Over the day, the quality of our decisions become visibly worse, but above all, the more decisions we make, the more we struggle to make others and the more stress we accumulate. Psychologist Roy Baumeister has shown in his numerous studies that it is useless to rely on willpower in most cases when we feel we are fighting against ourselves to remain clear-headed and effective: it would be enough to plan activities in advance so as not to go into energy reserves and to listen more to the signals of fatigue that our body launches us.[2] Baumeister explains how the smartest way around constantly making decisions is not to make them at all! How? By planning well in advance and creating functional routines for what we want to achieve, which make us act with minimal energy expenditure.

Beyond the Day

Energy management measures should be extended to the long term and to the short term: introduction of daily regenerating and "detoxifying" breaks, weekly, monthly, yearly, in which to detach from all electronic devices and work topics; balance and alternation between different work intensities; avoiding "dragging one's feet"; and introduction of "maintenance" moments to one's self-organization system.

Say NO

The world goes on without you. Resign yourself. Or rejoice.

Imagine you are in the office during a normal working day. Suddenly you get a call from home about an emergency, and you immediately have to leave the office and go. But on the agenda, there was a meeting you were supposed to attend and some e-mails you absolutely had to send. What do you do? You'll probably give a couple of tips on the fly to provide some relevant information for the meeting, send off just one of the urgent e-mails in less than a minute, without a lot of fuss and getting straight to the core, and leave to resolve the emergency at home. Or you'll just go away. The next day, or the day you return to the office, you will realize that everything went on without you. The meeting has been done. Your inbox hasn't exploded. Your company is still standing.

On the one hand, this is good news, because it makes you realize that there is ample room for flexibility in the management of what we often consider to be imperative with regard to our presence; on the other hand, it is bad news, because it explains the reasons why we desperately try to attend as much as we can in an attempt to prove to ourselves and to the world how important we are and how indispensable we are.

To achieve excellence, great results, and at the same time great well-being, we don't have to do everything, attend all the meetings, answer all the e-mails, all the messages, and all the requests that come to us from outside. We don't have to say yes to everyone. We don't have to accept all the proposals that come to us or do all the projects that we have the chance to be in. Paradoxically, we don't have to feel indispensable to anyone. **Learn to say no**. And figure out how best to do it.

Some cannot be yes moved in time: not now, but yes in 2 hours or in 2 days or in 2 months, considering our planning, perhaps negotiating dates, deadlines, and exchange of priorities with the interlocutor on duty.

Some can't be no forever: I will not accept a proposal for an investment or a corporate participation or a project opportunity.

Others cannot be spent on a more mundane level: meetings, business meetings, or evenings out that I can give up.

There are the no that I play on a thematic level: I decide that a whole category of actions goes outside my perimeter, such as the management of travels that I entrust to a secretary or other activities that I decide to delegate totally or partially.

There are some no we have to say to ourselves and some no we have to say to others. For example, to ourselves we might say no to the instant response to e-mails and phone messages—yes, you got it right: it is you who agrees to answer—or we might say no to certain micro-choices of ours, such as accepting all summonses or flying time requests made by colleagues and co-workers. At the meeting, your no could be replaced by the presence of a proxy co-worker or more simply by providing the information or opinion you requested for decisions involving your sphere. At flying time requests, your no could turn into periodic pre-scheduled time slots with colleagues and co-workers.

It is important to realize that when we have to say no, we enter a tunnel of guilt, qualms, fears, states of anxiety, which put us in front of ourselves even before those who have asked us for an answer. Should we say no to ourselves or to the other? The border is blurred. We must learn to say no to both, facing both our inner voice that urges us to participate for fear of losing importance, for fear of making a mistake, of being rejected, of missing an opportunity, and the other person outside of us with whom we must interact.

The next time we are asked for something, let us reflect for a few more seconds. We cannot control the inputs and requests that come to us from outside, but only the reactions we have in relation to them.

Let's focus more on ourselves, on the kind of response we really want to give based on the whole picture of the situation, bringing in the analytical and reflective thinking, which is a bit slower, rather than just the reactive, instinctive, and emotional thinking, which is faster, but not always able to choose the right thing for us.

Slotting

In business and in every professional field, it is unlikely to have only one project or one activity to be carried out in the time frames of our usual planning, from day to week, from month to year. In fact, in many cases, exactly the opposite happens. We are swamped by dozens and dozens of requests and are called upon to give output and answers on several fronts all the time. If we do not manage well the multi-projectuality of which we are naturally a part, we severely undermine our efficiency and face serious risks of burnout.

Imagine a large airport. How many planes land and take off on its runways? The busiest airport in the world is Hartsfield-Jackson Airport in Atlanta. Its five runways handle passenger traffic of around 100 million people per year. An airport capable of handling a total of 120 landings and departures per hour, or 22 per hour for each runway, approximately one landing or departure every 3 minutes. Do you know what the key rule is at this airport, which is the same as at all other airports around the world? It is called slotting, which is a disciplined system of planning and control, governed by the airport's control tower, which ensures that minimum interference times are respected between landings or departures, times necessary to eliminate the turbulence in the air created during a landing or take-off and indispensable to avoid any accident with absolute safety. And everything runs smoothly every day of the year at each airport thanks to this rule. All aircraft that want to land or take off will find their place on the runways if they wait for their slot. Simple. Without unnecessary stress and without the possibility of accidents. When we adopt the same slotting rule in the organization of our lives and our businesses, the first change is primarily at the cognitive level. We stop having that damned anxiety of having to give all the answers in the next half hour, of feeling drowned under the weight of all the tasks at hand, and we start being aware of the fact that the stress we experience is largely self-generated, that is it depends on how we react mentally and emotionally to all the things that happen to us and all the demands we receive.

Slotting gives us the cognitive tool to place everything in its place: in the day, in the week, in the month. We become the air traffic controllers of our commitment airport. Once you have placed the commitment, the activity, the project progress in its slot, the basic rule is: forget it! You will only deal with it in the moment dedicated to it, with as much concentration as possible. Perhaps in sacred time block mode.

In practice, if you have decided to have slots of 30 minutes each and if we consider a total active time of 10 hours, for example, you will have a beauty of 20 slots to "fill" each day.

Some slots can and should be in sacred time mode, some alone, some together with colleagues and co-workers, some more intense, some lighter, and even some slots will be pure relaxation, as we will see in the next countermeasure.

I decided to improve my work efficiency and that of my team to be able to invest time in development and stop chasing situations.

I could not find the time to see productions in the field to be more competent.

Through the Lean Lifestyle method, I set myself the goal of reducing the time I spent reading/processing e-mails and going to the plant every day to follow production.

I then defined four slots for reading e-mails and a time window to go to the plant, from 2.30 p.m. to 3 p.m. In addition, every Monday morning from 8:30 to 9:00 a.m. I set up a Lean Lifestyle development meeting with my team.

To facilitate the new habits, I have used 3 tricks:

- I close Outlook when I am not reading e-mails
- I instituted sacred time by notifying the company that every day from 10:00 to 10:30 our office is in a meeting, then I had to switch to alternating sacred time between team members.

■ I use a notebook where I write down each day's objectives and various requests, I only use Outlook reminders for important deadlines.

Results? I used to have 200 e-mails every day! Now I leave the office and no longer have messages that I have not read. It has not been a walk in the park, but it has been an achievement of which I am proud. I have a more constructive approach and perceive a great desire to improve in my team as well.

Lucio Fenati, Lean Coordinator Orogel

Sprint/Relaxation

When a good athlete trains, in all sports disciplines, they are well aware of the importance of recoveries between efforts both in a single training session and between training sessions. If the athlete lifts weights, for example, and if they want to develop explosive strength and power, they know that the recoveries must be more abundant between sets and in this case they can and must lift high loads, while if they want to develop hypertrophy and resistant strength, they know that the recoveries must be shorter between sets and the loads must be smaller. The athlete themself is well aware of the meaning of rest between training sessions and knows that only when the body is given the necessary rest, by not training at all or by doing things that are completely different to the previous session, will it be able to express excellence in the next training session. The body magically rebuilds all the fibers damaged during training precisely when it rests and especially during the sleeping hours. And with this reconstruction comes the athlete's muscle growth. Do you think you are different from this athlete when you are at work? Just because you don't sweat and use your muscles? You are wrong.

Research confirms the similarity between us and that athlete in my previous example. Our brain has the same recovery requirements between one effort and the next and between one work session and the next—the same, if not greater. In fact, the brain is the organ in our body that consumes the most energy of all. In fact, although it makes up only 2% of the body weight, it consumes 20–25% of the total energy absorbed by the body, or 300–500 calories per day, even when the body is at rest, not engaged in any activity other than the basic ones: breathing, digesting, and warming up. Most of the energy used by the brain is to make neurons communicate with each other, through chemical signals transmitted in cellular structures called synapses. Claude Messier, a professor of psychology and neuroscience at the University of Ottawa in Canada, explains that the more complex the activities are, such as learning music or engineering innovative moves in chess, the more energy we consume. On the other hand, we only must think about how we feel when we can be really concentrated on one subject for prolonged periods of time—studying, learning a new technique, reading hard: exhausted and in need of a break. This explains why **our days should be punctuated by real "sprints" of concentration and cognitive effort and healthy, rejuvenating relaxation breaks between sprints.** Breaks of 5–10 minutes are enough to give us the energy we need to dive back into our activities as intellectual athletes. And after a group of 3–4 sprints/relaxation, a longer break of 30–60 minutes will give us an additional energy boost.

As any good athlete knows, knowing how to manage recovery and rest periods well is crucial both for performance in the next training session and for the long-term sustainability of long and exhausting training programs. The same rule applies to all of us intellectual athletes. Any mistake or approximation in the management of our breaks and rests will have repercussions for both our short-term and long-term performance.

Someone must be thinking "but are we sure we need to take relaxation breaks at the office . . . and if I don't feel the need to take a break why do I need to take one?"

Beware, because if you really do not feel this need or if it seems so strange to think about it, it simply means that you are working at a low intensity, with low concentration, and tackling tasks that are relatively easy for us. This may well be the case, but surely in this case the results we get will be just as poor as our intensity.

In the company, the theme sprint/relaxation is something of a cultural taboo, because in most cases, people feel embarrassed to be seen "on a break" or "relaxing" in the workplace, except for the canonical coffee break with a possible junk snack at the vending machines. Or we only feel compelled to take a break when we feel exhausted from having spent several hours in unbridled multitasking mode with dozens and dozens of windows open in parallel, with our adrenalin and cortisol levels sky-high, tired, but ultimately with few results compared to what we could have achieved with less stress.

The more we get ourselves and others used to work in sprint mode, i.e. with very high concentration and focus, the more natural it will become to include relaxation phases at the end of each sprint.

It is the first time I have found such a visceral connection between performance at work and personal well-being. Now I feel empowered, and I am convinced that with my actions I can also have a strong impact on the people around me. I have acquired important tools. I have already seen results in the first month and today I am more and more optimistic: I had no more energy, now I am recharged and the goals I have set seem achievable.

Angela Grilenzoni, Quality and R&D Director EMEA Campari Group

Batching and Leveling

Would you ever think of going shopping every time you realize you need something at home? Milk is missing, I go to buy it. After an hour you realize there is a shortage of water bottles and get out again. As soon as you get back you feel like eating some fruit and not having it you go running to the greengrocer. Same thing for detergents or various utensils: as soon as the need arises you run to buy the product. I think very few people have such reactive and impulsive behavior. Generally, unless there are exceptionally specific needs, it is more logical to prepare a list of things to buy and divide them into one or two weekly expenses. Personally, I also try to concentrate on them in times of least possible affluence, for example the time slot when everyone is having lunch or 15 minutes before closing so that I am forced to be as quick as possible. If this is logical for one of the most common household tasks, grocery shopping, it is not clear why when we are at work, in the company or at home, we end up adhering to the complete opposite behavior. Incoming phone call? We answer, it seems obvious and natural. E-mail received? We rush to read it and reply. The boss claims us. Cannot wait, and then let's go to attention. And if a message arrives on the phone, we see it instantly because it may be good news. A voice WhatsApp? Not only do we hear it instantly, but we also "give" a nice little smiley reply (an emoticon is not denied to anyone!). But even when things get serious, a signature on a document, a request from a colleague, our answers are "yes, of course!" (5 minutes is not denied to anyone). Between these constant interruptions, surely, we don't want to deprive ourselves of a moment's pause to look at what's going on in the world? But, sure, then maybe the moment stretches a bit, and we end up browsing through online newspapers and videos of a football player's latest feat or a celebrity's gaffe on TV.

I could go on and on. The problem is not represented by all these small actions that suddenly drop in to interrupt whatever I am doing, but it is how I handle them. It would be much smarter and more efficient to merge all the little scattered things—which I often must do anyway—into specific, dedicated "batches," placing them in one or more focused slots. Example? E-mails, instead of being overwhelmed by them without a rule, read 15–20–30 times a day and at the most disparate and unpredictable times, you can decide to read them in dedicated batches. Once, twice, or three times a day. Stop. Interviews with co-workers? Batch. Depending on the intensity of communication I am required to have with them, I will batch daily or weekly or fortnightly or even monthly. The other small tasks that are necessary but with high interrupting power? Batching: signatures, authorizations, checks, messages, various errands, phone calls, etc. Yes, yes, you read that right. Phone calls. Even phone calls. Receiving and making phone calls is one of the most interruptive activities there is, because they project you into other people's worlds of which you have no control, unless you become almost rude. This is a good time to decide to run them in batching mode. Give yourself one or more slots dedicated to phone calls and merge them into those slots. A very good friend and colleague of mine has a habit of batching phone calls while jogging. I keep telling him that as soon as he switches to running, it will become impossible to continue with this habit, but he prefers to stay at a pace where he can breathe easy and talk on the phone at the same time. You can put in your batch of phone calls after an individual "sprint" with high concentration so that you completely switch cognitive channels and alternate the type of intellectual engagement.

But the final blow to inefficiencies from interruptions of small or large tasks will come when you combine the technique of batching with that of leveling.

What is it about? It is a lean production planning technique, *heijunka* in Japanese, which serves to level and balance workloads within production lines and minimize supply fluctuations. I personally learned to use this technique when I was responsible for logistics in the production department of a multinational company that manufactured automotive components back in 1996. Magneti Marelli, remember? I mentioned it in the first pages of this book. We had several customers to satisfy and our production, assembly and testing lines were no single-product and single-customer, but multi-product and multi-customer, that is when you needed to produce and ship specific products for different customers—almost every day—you had to stop the production in progress, change the equipment and set-up of the production facilities, and start manufacturing new products for different customers. This resulted in considerable loss of time and production efficiency. In the Lean world, we are "trained" to keep the so-called setup times as short as possible, and so we did in our case too, but below a certain technological limit it was impossible to go, which is why it was difficult to sustain a large number of setups and, above all, it was a source of real chaos to change without proper planning and preparation, both logistical and technological. But common sense is often far from good practice, and I still remember today the chaos we encountered when requests arrived from an important and well-known French car manufacturer, who never requested large volumes, on the contrary he used to make small requests, but when he did, he demanded what he asked for with absolute urgency, forcing us to change lines quickly with all the negative consequences you can imagine. The solution I adopted was precisely that of leveling, using a beautiful *heijunka* box,[3] with a time horizon of 4 weeks, in which I placed all the setups in advance, heedless of the fact that there was already an order from that customer. In fact, a stop was scheduled every 3 weeks to produce the quantities from the time series calculation and placed in stock. Somewhat contrary to traditional thinking, this method has the simultaneous objective of reducing variability, increasing the service level, and keeping stock low. Having six months' stock of a low rotation product would in fact cost us much less than having three days' stock of a high rotation product. On the other hand, no more chaos due to interruptions, logistical and technological unpreparedness, pressure from the customer. On the contrary, from that moment on, the orders of the customer in question would have been satisfied to almost zero time, taking the material from the stock already present in the warehouse, without any disruption to the production lines and logistics department. I made a different choice, however, for the products of so-called high-rotating customers: frequent changes, but well prepared and scheduled within the week, with stock tending to be very low, but frequently supplied, without large fluctuations.

Imagine applying the levelling technique to your personal or corporate group agenda: if you know you must do things, don't wait for them pounce on you when you least expect it, but plan with an extended time horizon, typically longer than a week, and already put commitments in your diary with different frequency. You will find that you have daily recurrences, while others will be weekly, and still others fortnightly or monthly or quarterly or yearly. Do you think it's too much? If you get used to it, you will appreciate the "calming and rationalizing" power of it. You will stimulate your adrenalin production through higher and more interesting challenges, as opposed to the chaos of random and unplanned commitments and tasks. Believe me, you will never come back.

Lean Lifestyle Story: Orogel

Small Solutions, Great Results

Interruptions, rework, stop-and-go and so on: the marketing department of Orogel appeared to be the last area of the company where working methods could be introduced to limit interruptions and incentivize highly concentrated work, reducing stress and the constant sense of urgency.

Instead, from 2020, it has been shown that a new way of working is also possible in a team where very different activities are managed (communication, marketing, digital, graphics, PR, and many others) and where people with different profiles and needs operate: digital specialists, graphic designers and press office.

Giulia Rossi, head of Orogel's marketing department, has acted in four directions:

- ■ Visual management/batching and leveling
- ■ Slotting/workload management
- ■ Sacred time/sprint-relax
- ■ Lean mailing

1. **Visual management/batching and leveling.** The projects that are taken over are broken down and scheduled into production phases. Once a week, in front of the boards set up with all the recurring activities of the year and those relating to the current week, the progress monitoring meeting takes place with the aim of aligning the team on completed and ongoing activities.
2. **Slotting/workload management.** The first part of each day is dedicated to particularly important tasks, those for which the most energy is needed, and particularly large tasks are divided into various slots to make the flow less heavy and ensure the right motivation to complete projects.
3. **Sacred time/sprint-relaxation.** People can freely choose when to enter the high concentration work mode by wearing headphones. This serves as a signal to office colleagues, while on the door to the workstations it is clearly explained to the rest of the colleagues that

the headset worn is the "do not disturb" signal. This has led to a huge reduction in mutual interruptions.

4. **Lean Mailing**. The reading of e-mails was limited to only 4 times of the day (09:30–11:30–14:30–16:30). Message management is based on a few simple rules: messages that take less than 2 minutes are processed immediately, while the others scheduled between the day's activities are postponed with the aim of not "reworking" the e-mails several times.

With just a few changes in the way we work, we have cleared the backlog, improved quality and reduced errors. Tasks that we used to work on for a whole week are handled in a morning and the energy put into individual tasks is much higher than before. We are less stressed because scheduling, alignment, high concentration work, have improved our well-being and urgencies are no longer the rule, but a task that finds a precise place in the work schedule.

Giulia Rossi, Orogel Marketing Manager

Reduce the Quantity and the Size of the Ongoing Activities

Imagine having to clean and tidy up a garage that you haven't tidied for years. Work tools, forgotten items, various paperwork and binders scattered around, some furniture that hasn't been used for years, dust a bit everywhere. How would you feel about having to sort out everything? And about having to do it in a single morning? A subtle feeling of paralysis would creep up inside you, leading you to reject the idea of having to do it. By chance, another task or urgency will become more important than that garage to be cleaned out, and that's it: procrastination until a date to be defined. You may even feel a little guilty, but you tend to find various justifications, including your supposed lack of willpower. Nothing could be more wrong. Willpower has nothing to do with this. The same thing happens with all those things that you want to deal with in batches that are too big for your brain to accept. And especially in situations where your brain struggles to see the actual conclusion and its final output. When this happens, when the image of the result is too far removed from the actual reality, we enact silent acts of self-sabotage, because we want to avoid the enormous effort needed to overcome that gap.

Everything changes when you can see the conclusion of your efforts. If, instead of the whole garage, you were to clean only a small part of it, well-identified and requiring part of it in a short time, magically you would meet less resistance and most likely bring the task to a conclusion. Perhaps it will then be easy to proceed to clean up another piece on the same day or another, and so on.

The secret why this mechanism happens is explained in a beautiful article published in 2011 in the *Harvard Business Review*, "The Power of Small Wins," in which the authors reported the results of observations made on hundreds of individuals and teams, from professionals to research teams and from industrial technicians to famous inventors. The data collected showed that one of the most powerful weapons in the most successful and innovative individuals and groups is the experimentation of what they called a "sense of progress." The more you get used to experiencing small breakthroughs, the more you feed the emotional side of individuals and groups, which become more "creatively productive" over the long period.[4] In essence, if we see small tangible results from an activity performed, we are stimulated to do it, to redo it and to persevere because the resulting element of emotional gratification acts as an engine that feeds the action. Imagine a researcher struggling with his own lab activities that are long, tiring and often lacking in striking results. What will make this researcher persevere? Or imagine an athlete training hard for any competitive sport, day after day, week after week, and month after month. What will push this athlete to keep going? In both cases, certainly not the tangible elements from the outside. **Motivation often comes from a series of seemingly intangible elements**, in particular the search for and obtaining small signs of progress; sometimes they are very weak signals but able to keep our "internal engine" running, which can change the meaning and significance of all the actions we are able to perform. It is not a rational mechanism, but one that can positively affect our emotional state by providing us with energy we did not think we had. In the case of the researcher, for example, the elements capable of keeping the internal emotional engine running, even for a long time, may be a small sign of evolution in one's daily laboratory activities, together with the clear vision of a deeply desired result and the climate of confidence around him or her. For the athlete, on the other hand, these signals may be having improved a test by a few tenths of a second or having lifted a few extra kilograms, together with the winning self-image in a competition, even if months and months away from the present moment. In both cases, it is essential to note another strategic element: both the researcher and the athlete "feel good" when they experience this sense of progress. And the better they feel, the more productive they will continue to be in their activity. All of us, however, suffer from a small cognitive bias that we can actually exploit in our favor: our brains do not distinguish well the difference between a small and a large task or between a more important and a less important one. When we successfully complete something, we are happy, and a small rush of dopamine makes us feel fulfilled. But success calls success: when you have completed one small task, you feel like finishing the next one, entering a real virtuous circle. Your brain thanks you, your emotional state is invigorated, your mental energy increases, and things flow.

In essence, we will find great benefit whenever we "break" a large task into such small quantities that we can consider them terminable in small portions of time.

I completed the writing of this book in a particularly difficult period, both for reasons external to me and due to personal events, that occurred between April and May 2020. The Covid-19 pandemic from February 2020, has forced us into lockdown periods, restrictions of various kinds, and all the other consequences that we have come to know well, from the unknowns in terms of our health and that of our loved ones, to the unknowns in terms of employment and the effects on our businesses. On a personal level, the heart attack I suffered changed my lifestyle and accelerated some individual and business decisions. I have never had as many "headaches" as I have had in the last few months, both on the professional and family front, but the dream of being able to produce my new book and perhaps publish it on the "restart" for all of us was too big not to be cultivated. Great was the dream of spreading the Lean Lifestyle principles at the very time when, in my opinion, we needed them most. But also great was the gap to be bridged in the short time available. I exploited all the principles seen so far: 1440 mindset, selective silence, sacred time blocks, energy management, say NO, slotting, sprint/relaxation, batching and leveling in addition to this to this of the reduction in quantity and size of tasks. Each day I give myself several slots of 2 hours each, in which I just write. But in each time slot, I don't write the book. I write a single page, I deal with a single principle, I delve into a single technique. The 2-hour slots are broken by a small 10-minute break in which *I enjoy the output I have just baked.* It can be a half page or a full page. And I am only focused on this next small output. In this way I did not feel the weight of a whole book to write, which would have sunk me and pulled me away from the action. In doing so, I moved forward nourished by a sense of daily progress. When I sit down, I feel in anticipation the thrill of the next page that I will see realized before long. I try to pay attention to an important psychological element. The sense of progress does not manifest itself in things left half-finished, things unfinished. When I finish this piece of the book shortly, I will have finished and realized this piece of the book and that's it. I will enjoy it. I am neither a third of the way through the book, nor half-way through it, nor even with another 100 pages still to write. The feeling is completely different. If I want to keep my emotional engine running, I must focus on the page to be completed and see it finished after one or two hours, rather than on the unfinished work that is still to be done. At the end of the written pages, and of the other slots scheduled for the day, I gave myself the best gifts of the day: the sacred time I allowed myself with Francesca, my wife, with whom I had lunch and chatted, forgetting all trace of work, or with Amedeo, my 3-year-old son, with whom I played and had fun during the time dedicated exclusively to him.

The quality of life depends on the quality of the individual moments we live every day.

One or Two Top Priority Every Day

Imagine taking part in a target shooting competition. What would happen if the number of targets to be hit suddenly increased at the same time, allowing you to choose what to aim for in the competition? Most likely you would lose focus, you would not be able to concentrate on the single target to be hit, and at the end of all the number of targets would be drastically lower than a situation where you could have focused on one at a time.

It is normal that our brain when faced with so many possible targets to hit almost goes crazy, constantly changing focus to help us hit some. We experience this dilemma every time we start our working day. Don't let your seemingly already well-organized schedule fool you either. The dilemma first plays itself out in your mind. If you have not done the preparatory work—before you start your day—of focusing on priorities, you are very likely to scatter yourself into dozens of rivulets, tasks and fake urgencies, perhaps other people's priorities, ending up opening many windows and closing far fewer than you open.

Why Only One or Two Priorities?

We are much more experienced in opening new fronts than in closing them. In fact, each of us has dozens and dozens of open and pending fronts that have accumulated over the years. Look carefully at yourself.

Now take a few minutes to make a list of all the things you would like to finish. Put in small things, big things, work, family and personal. You won't struggle to compile a list of dozens and dozens of things still unfinished. The point is that all these unfinished things are constantly surfing through your mind in the "background," playing the same trick on you as the tug-of-war mentioned above.

The Lithuanian psychologist and psychiatrist Bljuma Zeigarnick demonstrated this theory in the last century and formulated the so-called Zeigarnick effect. The story tells[5] that the studies on this theory were initiated following a curious observation made by the psychologist herself in a crowded Viennese restaurant, in which a waiter remembered very well and without the aid of any written notes all the orders that had been partially fulfilled, while he remembered nothing of the orders that had already been completed. Intrigued by this observation, Dr. Zeigarnik started a series of tests in her laboratories on a group of participants to prove the theory she had hypothesized to explain this waiter's strange behavior. Participants were asked to perform about 20 simple tasks: solving puzzles, making necklaces, playing games, solving riddles and arithmetic problems. A small detail of the experiment: the participants were often interrupted during the execution of their tasks and the exact point at which the tasks were interrupted was traced. At the end of the experiment, the participants were asked which, of the 20 or so tasks performed, they remembered best: almost all of them remembered twice as many unfinished exercises as successfully completed ones, exactly as in the case of the waiter observed in the Viennese restaurant.

The Zeigarnik effect demonstrates that when a task is not completed, a mental state of tension is created that prevents the mind from effectively starting another task from scratch. It is the same mechanism that is put in place when a to-do list is drawn up or when putting a task on the agenda. Just because you have planned an activity, the mind sends anxious messages urging it to complete it, preventing you from concentrating on other mental processes. It follows that, to forget and stop thinking about something, whatever it may be, it is indispensable to "finish" the mental action begun earlier.

This effect is also exploited on television, when a series is interrupted in the middle of an action or story, creating so-called cliffhangers, which leave viewers' minds eager to see how the story ends in the next episode. The same principle is followed in cinematic video trailers, leaving scenes deliberately cut for the same reason or in films that in the cinema have made use of this narrative element: for example, the *Harry Potter* and *Avengers sagas*.

It is for the above reasons that it is useful to insert one or two top priorities per day, or one or two activities to be completed completely, taking care never to include too many different things in the same day. Imagine a week in which you concluded 7–14 top priorities. A month with 30–60 completed top priorities. A year with 365–730 completed top priorities. Maybe even too many. Aiming for 200–300 top priorities to be completed in a year would be a great achievement.

When to think about the priorities to be completed in the day? I suggest doing it in the evening, so that we already have a clear picture of what the next day has in store for us as soon as we wake up and prevent other non-priority things from creeping into our heads before our agenda. During the night, our brain will process in the background the information received about the next day's priorities and this will help us to focus better since waking up.

Lean Lifestyle Story: Garavini

Growing in Times of Crisis

Small, but ready to grow up. In Garavini, a company in Forlì specialized in the production and sale of upholstered furniture and furnishing accessories for the Italian and French markets, the Lean Lifestyle has revolutionized the way of working.

> "The organization of our work, the focus, the routines for monitoring and controlling production and the budget," says owner **Vanni Garavini**, "allowed me in a year, 2020, which was so complex and uncertain, to keep control of the situation, not to lose focus on the company objectives and to act quickly with the necessary course corrections, thus closing the year with a 22% increase in turnover, in a sector, such as that of furniture, where it seemed impossible to see a positive sign. I did not think that such structured working methods could work and be sustainable in a small company like ours."

Also, for **Katia Rigucci**, general manager, the change of pace was evident:

> My way of working has changed and today I can say that I start a day in the company with a serenity and energy never had before. This has been possible thanks to the acquisition of new

individual and organizational habits aimed at focusing our work on value-added activities and creating regular routines in which I can perform my job of overseeing production and controlling the budget more effectively. The Lean Lifestyle method has reduced wasted time and helped the company at all levels to work with never before experienced planning. We have reduced stop-and-go because communication and control moments are clearly defined and shared. Everyone knows their respective windows of availability and we respect our colleagues' "sacred time."

C.E.P.E. (No to Do List)

One of the world's best-selling books on personal productivity is American David Allen's GTD *Getting Things Done*. The method described in this book has been used by millions of people worldwide and adapted in various forms by probably even more people. Published in its first edition in 2001, the book marked the birth of a new way of thinking about one's personal organization, but perhaps today it finds its limitations in the excessive length and "bureaucracy" of the method, in the absence of digital tools to support it, and the excessive use of paper and physical elements that are, in my opinion, outdated.

For these reasons, I propose a useful summary, which overcomes the aforementioned limitations and simplifies and facilitates its use while keeping its basic principles valid.

The Logical Levels

I start with a principle to clarify, before going into operational details.

Attention to different logical levels, or to the different cognitive missions. In my opinion, the concept of *time management* is wrong from the start. You cannot manage time in any way. As I have already repeated elsewhere in this book, time is that and it is the same for everyone. You cannot manage it. You can't lengthen it, you can't shorten it, you can't distribute it, you can't put it in a safe, nothing. You cannot give it any order. The only action you can take is to manage what you decide to do with the time you have. One of the main problems that makes us mentally go haywire and severely damages our effectiveness is the overlapping of different logical levels at the same time: is the clumsy attempt to multitask, or more correctly rapid task switching, between different cognitive objectives while apparently remaining in the same place. If I am performing a task, I cannot simultaneously plan what I will do next or tomorrow or in a week's time. If I am planning, I cannot perform any specific task. If I am focused on a task, I cannot simultaneously collect all the inputs coming to me from the outside world or from a remote corner of my brain in Zeigarnik anxiety effect. I will struggle to process the information I got and understand what to do with it if I am doing something else. We, on the other hand, often embark on convoluted and mentally unproductive cognitive quests daily, not leaving the right amount of space for each of the logical levels just described, which need their own mental space to be carried out at their best. Here are the logical levels to distinguish:

1. **collection of** inputs that we receive continuously;
2. **input elaboration** and decision-making on each;
3. **planning of** the agreed steps for each elaborated element;
4. focused **execution of** planned tasks.
   Collection, elaboration, planning, execution: C.E.P.E. (Figure 5.1)

1. Collection

Let's start with the logical level that I have called "collection." This is a level that should be left to act in the background, taking advantage of real multitasking in one of the few circumstances in which it is really useful. Right now, I am writing and in parallel the collection tank of my mailbox is filling up. And I let it fill up in peace, just as WhatsApp, LinkedIn, and Facebook message catch basins are most likely filling up. I let them fill up in peace. And since my phone is in silent mode, out of my sight, the catch basin of unanswered phone calls must be filling up. I go even further. If from a remote corner of my brain comes an idea or an alert for something to be done, no problem: I will jot it down—without thinking too much about it—in a notebook that I always carry with me. The alternative to the notebook is the TO DO application on my phone, which I do not use when I do a highly concentrated activity such as writing, because I want to avoid the very high risk of distraction that I run as soon as I touch the phone. Many people turn up their noses in front of the word distraction, but that is precisely what distraction is: distraction from the main task on which you have previously decided to focus. Needless to say, that in this mode I have all the ringtones of the various notifications completely deactivated and I use a

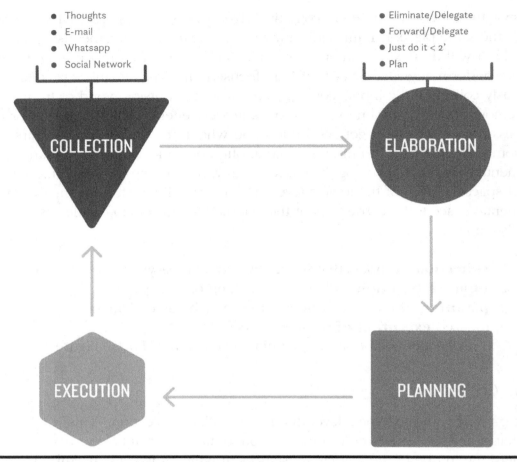

Figure 5.1   The phases of the C.E.P.E. system.

Microsoft Office function called "focus mode" that allows the computer to display only the page I am writing on and nothing else. Exactly as if it were a traditional typewriter. All this is "accepted" *by* my mind due to the fact that I can "trust" the system that has been built, which provides for the simple, yet structured collection phase, followed by the next phase, that of processing the information received.

2. *Elaboration*

In this phase I open one by one the containers where information and input have been quietly accumulated for me and I decide, one thing at a time, what to do. In this valuable and focused processing phase, I will be able to make several decisions:

   i. unscrupulously eliminate the input or information received, because it is considered unimportant to me now;
  ii. forward to colleagues, friends, collaborators if I believe there is an interest in them;
 iii. delegate to those who can and must carry out an activity linked to the input received;
  iv. perform the task directly if and only if it is short, generally under two minutes;
   v. plan in one of the future available slots, when to perform the activity linked to the input received that I feel I must do;
  vi. erase all traces of input and information received after my processing!

How many times do "elaboration"? It depends on the type of work each of us performs. Generally, I associate the processing phase with the daily moment of reading e-mails, the frequency of which should never exceed 3–4 times a day, unless you are a call center operator or at the front office of a customer service department or other similar jobs. Generally, with a daily frequency of 1–2 times you survive very well and optimize your personal organization a lot.

3. Planning

The next phase, planning, is precious in the C.E.P.E. (Figure 5.1) scheme because it becomes a constant element to refer to during any processing, or during any moment of indecision as to what to do: it will give me the lucidity necessary to understand whether to accept and when to accept new tasks assigned by myself or arriving from outside, it will give me the certainty of being able to verify whether the objectives I would like to achieve are more or less likely to be achieved, because looking at the agenda I will quickly understand whether there are activities scheduled useful to their achievement. Scheduling becomes a ritual of checking the whole system, a delicate and intimate moment of self-reflection in which to look back at what has been done, asking questions about how and what can be improved in the next scheduling period. I generally use a weekly slot of 1–2 hours dedicated only to planning. And I try to enjoy it all, reflecting on what went well and what went less well in the weekend. I check whether my plans have progressed and whether I need to "fix" something in my system. Over time, I have learnt to appreciate these moments because they are real "pit stops" that allow me to start again more lucid and focused than before. **To the weekly planning, which I do with a time horizon of 3–4 weeks, I follow up with short reports at the end of the day in which I review on the fly the outcome of the day ended and calibrate the next one by focusing on my top priorities for the next day.** Over the years, I have developed a visual system with different colors on my agenda in Outlook, which gives me briefly the size of the various areas in my life. For example, the strategic gold activities for me are deep green, sports activities light green, and family activities olive green, while client activities are blue, business development and growth activities red, administrative, and human resources activities yellow, and those where I do online or live speech dark gray. Since May 2020, the golden color has become clearly visible and distributed, the one dedicated to my health and deep well-being: meditation sessions, flexibility and yoga routines, medical examinations, taking care of my body.

With a simple view of my calendar, I see if the balance matches to what I set in the annual planning session, the one in which I take stock of the past year and plan the following year. This macro-session usually lasts a couple of days and is followed by shorter review sessions—about three or four hours, another two times during the year.

I can already hear you: "Too complicated Luciano, we seem to fall imprisoned in a system that involves discipline and rigor. But like all habits that you learn, once you get into your pattern of life, I assure you that it all becomes simple and above all you enjoy all the benefits that come from having just patience and discipline. The biggest benefit, for me, is that I really feel free to live the life I want and really achieve the goals I set for myself. Many people, on the other hand, remain slaves to their own beliefs and their own non-system of self-organization in which they are trapped, believing to be lighter and carefree.

I only know one thing about this: you only realize things that have been consciously planned. And since all planning is an act of trust in the future, it is necessary to continuously monitor whether your intentions have been followed by concrete action or not.

4. Execution

All this is necessary to jump headlong, free, focused, and with incredible energy onto the fourth and final phase of the C.E.P.E. system: the execution of what I have chosen to do. When I am in this phase, protected by every-thing I have solved in the other three previous phases, believe me, it is a whole different story. The famous "here and now" becomes reality. My mind really becomes like water.

Pull Planning

Imagine you going to a restaurant. As soon as you arrive, you are offered a place to sit, your coat or jacket is adjusted and as soon as you are seated, the waiter starts with questions: do you prefer still or sparkling water? May I offer you an aperitif? Alcoholic or non-alcoholic? You are left with a menu and after a few minutes you are asked what you prefer to eat, also receiving additional information and explanations about what is on the menu. Based on your choices, your orders will be taken to the kitchen and the team consisting of the chef and his staff will start prepar-ing your chosen courses. It would be unthinkable for dessert to arrive first and then the first course. Quite simply, the restaurant and its kitchen have carried out a plan "pulled by the customer's needs." The raw materials and at most some semi-finished products had been sourced and arranged before your arrival, but the cooking and final preparation of your dishes was done according to your wishes and only for you. The principle of Pull Planning is very well known to those involved in the application of Lean Thinking in production areas, where there is a fundamental rule: you do not produce what has not been requested by a customer downstream in the value chain, whether internal or external to the company, except for the creation of small safety stocks between stations or in the warehouse, just to cope with fluctuations and excessive variability in demand. The just-in-time technique involves having the parts in the right place, at the right time, in the right quantity in a production system and is based on the various applications of the so-called kanban for its implementation: physi-cal or electronic tags or containers that moves from one work station to another, from downstream to upstream, as production signals and autho-rization to restore intermediate stocks at downstream stations. Returning to the restaurant example, the sheet or tablet on which the waiter marked your order represents the production kanban for the kitchen, which is only authorized to produce the quantities indicated in that specific order. By relying on these principles, many companies, starting with Toyota, have been able to quickly meet the needs of their customers, while optimizing their production-logistics chain with the minimum number of stocks, materials, and finished products in their warehouses. Well, if for a restaurant is known and sensible thing, if for a lean production system, it is synonymous with agility and efficiency, in the office and intellectual work environment this principle is not at all familiar, either because of the frequent lack of knowledge of it in that sphere, or because of resistance of all kinds to its application. Being guided by what a customer downstream has requested and to realize only what is necessary is a guiding principle with a very high impact both at the individual level and in teamwork, especially in the development and design of new products, software, services, construction sites. Jeff Sutherland, one of the creators of the Agile Work Manifesto, states that if this principle was applied extensively in offices, it would halve lead times and double output while drastically improving the quality.[6] Adopted by many of the most important companies in Italy and worldwide—from Amazon to Google, from Apple to Ferrari, and even small companies—Scrum is a working method that translates just-in-time concepts from production environments to development environments. One of the founding principles is the division of the entire software or product to be developed into small parts, and then entrusting the development of the individual parts one at a time to a small team of people. This team is called upon to realize only what pertains to that small part of the overall project entrusted, in a short period of time—called a sprint—generally of 1–2 weeks, under the guidance of a team leader who plays a key role in representing the voice of the end customer, ensuring that there are no external disruptions and that the team focuses only on that task. At the end of the sprint there is the validation of the output provided with the participation of the management, the customer's representative, and sometimes the customer themself. Only after these rapid and frequent intermediate validation sessions do we proceed further, with the application of any corrective actions or the development of the other parts of the project. The fundamental aim of the Agile or Scrum methodology is to avoid at all costs the development of elements that are not required by the customer or not deemed to have added value for him, optimizing the use of resources and minimizing waste.

Among the many lessons learned by Jeff Sutherland, one in particular will benefit you enormously both individually and as a team: let yourself be guided only by what you want to achieve, specify it well, break it down into small parts and plan accordingly. And if you are working in a complex context, where other colleagues expect something from you, remember to apply the Pull concept, which is to get pulled by what your colleague needs.

Tip: ask and compare what is more functional to be done first. Next time, for example, you go to a meeting, be ruthless in focusing on the output you want to achieve, the decision you want to make, and steer the meeting discussions all in that direction. Surely in any business or professional environment you will do things intended for one or more customers, internal to the company or external. Get into the habit of frequently asking what these customers expect from you and frequently checking with them to see if you are going in the right direction or not. All this means acquiring an Agile mindset, typical of people who do not move by doing scattered activities, but first plan the results they want to produce, and then the actions to get them, quickly and without waste.

Lean Lifestyle Story: Danieli E ABS-Acciaierie Bertoli Safau

The Toolbox

A "breath of fresh air," as simple as it is vital. The first steps in adopting new ways of working on the road to the Lean Lifestyle are those that, in their simplicity and ease of implementation, give the first great satisfactions and, above all, the energy not to stop and not to turn back. This is what happened to **Alessandro Trivillin**, former CEO of ABS Acciaierie Bertoli Safao and the Danieli Group, and currently CEO of Snaidero Cucine, when he decided to use the Lean Lifestyle leader's toolbox.

"I felt like the famous cartoon character Handy Manny: I opened this box and began methodically applying the tools on myself, my co-workers and the entire organizations I was leading. First, I asked my secretary to put some empty slots between appointments so that I could have buffers and be able to apply sprint-relaxation logic," he recalls. "In addition, I told her not to make any more appointments before 9.30 a.m. because until then I would always have sacred time for my gold activities. I gained great serenity because I eliminated the feeling of oversaturation and gained time to focus on value activities daily. It has not been an easy path because the temptation to fill in the 'blanks' remains, but the way of working has definitely changed: I am no longer guided by events and external demands in setting my agenda, slotting and regular cadences govern my actions. In this change, I got help from my co-workers and I constantly asked for feedback, so that I could refine my improvement project and understand better and better what my Gold Activities were. In this process, I came to reduce some activities and delegate others, always making sure to explain the delegated task very well."

Among Trivillin's new ways of working is of course the e-mail management chapter, an activity that he admits "has always given me a dopamine rush."

I am not yet immune to the constant attraction of unregulated reading of messages, but an important step to reduce multitasking was to eliminate email notifications and to give myself dedicated slots. I have defined some people as "VIPs" for whom e-mails come to me as notifications, all the other process no more than once an hour. Then there are some cases where I turn off all notifications, even the VIP ones, such as during meetings and One-to-One with people. And finally, I also deleted all unnecessary newsletters and put in a series of filters that cleaned up the inbox.

Some advice?

Don't be afraid of dimensional constraints: these tools can be applied by an individual, in a family (which is what I did!), in a company of 1,500 people and a group of 10,000. The difference is how to put them on the ground.

Establish Regular Basis for Thematic Areas

The role of habits in our lives is fundamental. As I mentioned in Chapter 2, I believe that our life is determined by two things, the decisions we make and the habits we establish. The former gives the direction, and the latter determines the person we will be along the way. What you do every day and how you do it shapes you to the point of making automatic many of your behaviors and seemingly spontaneous gestures that you have acquired and repeated so many times that they are performed without even thinking about it anymore. Everything we have discussed in this chapter is about possible new habits functional to optimal planning and our goal of gaining precious hours at the end of our working days. The techniques we have seen so far find their best realization when they form part of an orderly and regular flow of thematic activities that alternate regularly with each other. I can make use of sacred time, slotting, energy management, pull planning, and C.E.P.E and certainly gain tangible benefits in terms of time gained and stress reduction, but I will gain much more when I harmonize these techniques into my own system of regular cadences in which the various activities in my personal and professional life follow one another. Regular cadences can be set for the single day, week, month, quarter, and in the year. The objective is to minimize interruptions due to changes of context or project, in the work and personal sphere, with the aim of making the most of the boost in concentration that we find when we focus only on one topic, forgetting the others. Another goal, trivial, but that will earn you many hours in planning is to plan in advance many of those things that you would otherwise do every day and every week starting from scratch all the time. And I am not talking about planning today what I will do in detail in 3 months' time, but creating a structure in which it will be much easier to do in 3 months' time. Let's see some examples. In the day I will have benefits when certain slots are always done at the same time: morning routine, evening routine, lunch, dinner, workout, e-mail management, and sacred time for daily top priority. In the week or month, I will have even more benefits, even more when I set themed days or half-days for individual work areas. Dan Sullivan even suggests dividing the days of the week into three types: Buffer Days, Focus Days, and Free Days. In his *Dan Sullivan's Entrepreneurial Time System®*, time is structured into three very different types of days to ensure maximum performance and still plenty of free time for personal interests. Many successful people use this system to achieve great results and maintain a balance between work, family, and leisure in their life.[7] A Focus Day is a day where you spend at least 80% of your time leveraging your personal talents or working on an area of primary competence for your business. Depending on your profession and your talents, your Focus Day could be committed to your defined Gold Activity that will give you the highest return on your time invested. Buffer Days, on the other hand, are days spent preparing for and eliminating distractions during a Focus Day or Free Day. Buffer Days are important because they ensure that your Focus Days are as productive as possible. The key is to group all Buffer Day activities on the same day so that you do not dilute your Focus Days and Free Days with activities that would distract you from your main task or prevent you from really enjoying your free time. In a Buffer Day, therefore, you solve the issues that could take away concentration during the Focus Day. A Free Day is conceived as a full day that is completely free—from midnight to midnight—and has no work-related activities of any kind. It is a day without business meetings, work-related phone calls, e-mails or reading work-related magazines and documents. On a real day off, you are not available to your colleagues and co-workers, customers, or students for any time of contact except for real emergencies: accidents, deaths, floods, or fires. These days off are vital because they allow you to relax, have fun and spend quality time with your family and friends, so you can return to your work regenerated and ready to deal with it renewed vigor, enthusiasm, and creativity. Perhaps you'll find that quite a few new and creative ideas will come to you because of these regenerating moments that you put into your week.

How can you apply this powerful principle of regular cadences by subject areas to dramatically increase your productivity, even if you do not have complete control over your agenda?

- You can choose a day of the week to hold all recurring internal meetings. On that day, for example on Monday, you can hold an individual meeting with each of your direct collaborators. In these "1 to 1 meeting" you can assess what happened in the previous week and priorities for the new week, review key metrics, and give support and contribution. This will save all pockets of time lost to miscommunication, keep people focused on their weekly responsibilities, and eliminate all those flying "do you have a minute for me?" meetings. Which are usually never a minute.
- You can choose one day a week to ban all internal meetings (at least on your work team) and have people work in "sacred time" mode on their key projects.
- You can choose one day a week for all administrative matters, such as filling out expense reports, completing audits of all accounts, reading reports of various internal committees, recovering old e-mails and schedule any meetings that are not directly related to your key goals.
- You can choose one day a week to focus on your customers (both internal and external). Try to leave your office on this day. Make some phone calls, take your best customers to lunch, and listen to the feedback they give you.

■ You can choose fixed days or portions of the day, where you can always carry out the Gold Activities you have defined as high impact for you and your business.

Current managerial models are outdated because they do not really impact people's lives. With the Lean Lifestyle, I have put my professional and private life in order.

Michele Castiglioni,
Supply Chain & Logistic Manager at Stanley Black & Decker

Notes

1. Tony Schwartz, *The Way We're Working Isn't Working: Fueling the Four Needs that Energize Great Performance*. Free Press, 2010.
2. Roy F. Baumeister and John Tierney, *Willpower: Rediscovering Our Greatest Strength*. Penguin Books, 2012.
3. For more on this planning technique you can use the book: J. Liker, *Toyota Way. 14 Management Principles from the World's Greatest Manufacturer*. McGraw-Hill, 2004.
4. Teresa M. Amabile and Steve J. Kramer, *The Power of Small Wins*. Harvard Business Review, May 2011.
5. P. Banyard, C. Norman, G. Dillon and B. Winder, *Essential Psychology*. Sage, 2019.
   D. H. Gomes, *35 Mental Techniques (and Curiosities): Why the Mind Must Evolve Too*. Babelcube Books, 2020.
   Benjamin B. Wolman, *Contemporary Theories and Systems in Psychology*. Springer US, 1981.
6. Jeff Sutherland, *Scrum: The Art of Doing Twice the Work in Half the Time*. Random House Business, 2015.
7. For further reading: Dan Sullivan, *The 25-Year Framework: Your 21st-Century Entrepreneurial Mindset for Continually Slowing Down Time While Speeding Up Your Progress Over a 25-Year Period*. Author Academy Elite, 2019.
   Dan Sullivan and Catherine Nomura, *The Laws of Lifetime Growth: Always Make Your Future Bigger Than Your Past*. Berrett-Koehler Publishers, 2016.

Chapter 6

The Secrets of High-Impact Delegation

Delegate, Delegate, Delegate

When it comes to leadership or management, this is definitely the first thing you will have been advised to improve in these fields. Unfortunately, when you go and actually see how various managers or entrepreneurs, you realize that the 99% of the time the problem is not that they do not delegate, but it is that when they delegate, they do it downright badly, because they do not know how to do it effectively, even when they understand what they have to delegate.

In a project a few years ago, a client of mine, at some point, interrupted the activities he was doing with me by saying, in a rather firm tone:

> Luciano I have understood what I have to delegate, I have started to use delegation, but nothing has changed! I haven't seen results, in fact I've slowed down my performance because if before I used to take a certain amount of time to do things, now I even take twice or three times as long because I have to explain things. People then start doing things on their own and don't do them as they should. At this point I went back because maybe the problem is with my collaborators: they are not the right ones. I have to find different collaborators if I am to continue to exercise delegation as you have asked me to in this program. Let's suspend everything for now.

What did I do? I went to the field with him to see a co-worker and asked him to give me *a chance* to see how he was delegating. I realized that

DOI: 10.4324/9781003474852-6

the problem was not in the activities he was delegating, but in the way he was delegating. A series of fundamental errors prevented him from using delegation as an effective tool both for the development of his skills as a head coach—we will see what that means in a moment—and for the growth of his people.

The fundamental difference between an excellent delegator and a mediocre one is the ability to put their collaborator in a position to start the work, make him/her as autonomous as possible, and support him/her through effective supervision until the delegated work is completed. A delegation without good supervision has little chance of bearing great fruit. So, if someone told you that you can delegate everything you don't like about your work and forget about it, I have bad news for you: it's not true. Or at least I have never seen it happen in over 25 years of work and hundreds of organizations of any size.

A well-done delegation will certainly relieve you of so much weight that today weighs on you, but if your focus is only on your "relieving" yourself, you will never become an excellent delegator. This sounds like a paradox, but it is not at all.

Lightening is only the side effect of a high-impact delegation—which, in turn, it has its roots in the kind of leader you are or want to become— because only when your employees have grown will they relieve you of so many tasks that weigh heavily on you today. This type of delegation is not easy to implement, but it is the most powerful for you, your people, and your company, in terms of developing the human potential available.

Fortunately, there are some essential "techniques" to become truly effective and able to increase the impact of your delegation. And these techniques have now been empirically validated by the experience of hundreds of managers around the world. We can group them into five subsets:

1. what to delegate;
2. what not to delegate;
3. how to delegate;
4. to whom to delegate;
5. when to delegate.

What to Delegate?

There are two problems that hinder the growth of any manager and entrepreneur. Summarized in:

- "How do I free my vital and strategic time?"
- "How can I grow my staff?"

There is only one answer to these two questions. It is called high-impact delegation.

The first step is to **choose well what to delegate**.

Every day my staff and I are faced with the same dilemmas as thousands of managers and entrepreneurs in companies struggling with daily difficulties: To grow people, to free our precious time, to conduct an effective delegation mechanism.

The first remark I often get is: "My people are not capable of doing things the way I do them: what can I do about it?" The first answer I give in these cases is confirmatory:

It's true, you are right! There is no one in the world capable of doing things exactly as you do, and you will never find it!

But then I add a second answer that is the first step to effective delegation, and it is the following:

Delegate everything that your co-worker could do in your place, with a return of about 70% on you. The performance I am referring to must always be considered downstream of an adequate training that the delegate will have received from the delegator! So, stop now, think hard, do not go on reading: is there someone who could do at 70 per cent performance, compared to you, what you do today? Maybe after coaching by you or someone else who can mentor that person? If the answer is yes, you are done looking: delegate now to someone who can guarantee you this 70% and concentrate on something else!

What else can you delegate? Let's go even deeper. Some things to delegate are easy to guess, others are really counterintuitive.

a. Everything you can do so well that you can teach others, freeing up valuable time and energy for new challenges.
b. Everything that is not gold activity for you.
c. Everything that makes the collaborator grow.
d. All problem-solving activities that employees find themselves in forehead.

Everything You Can Do So Well That You Can Teach Others

Very often people apply delegation as if it were a blame game: I delegate what I can do less or that bothers me more, or what I don't really do well. And this may even be understandable and sensible in many cases, but the exercise of delegation becomes truly powerful when you decide to delegate what you do best. The effective delegator will paradoxically delegate the activities he is best at, the ones he really masters. And this will enable him to easily teach them to others and, above all, allow him a quick and agile follow-up. Why is it important to delegate what you do well? Because delegation is revalued, and it becomes a tool for people's growth. If I delegate something I am good at, at the very moment I am delegating I become a teacher and mentor. Delegation, in this case, also takes on a different function for the employee, almost a function of training on the job, learning in the field. In addition to having assigned the task to be performed to the person, I am also carefully choosing that task from among my best things, among what I have learnt to do best over the years, and thus I will be able to support my co-worker, teach him and give him valuable feedback.

Everything That Is Not Gold Activity for You

In Chapter 3, you learnt to identify the Gold Activities, those that are able to bring you the highest impact, with the greatest passion, on which you put your skills, and those that represent a real challenge for you. And it is from this precious personal list that it is easiest to start identifying what you can delegate, "fishing" in the complementary list, those of all the other "non-Gold" activities. Obviously after trying to reduce or eliminate as much as possible.

If there is something that does not work or is of little use, you must eliminate it and not delegate it. And if this step in identifying what you can delegate is rather intuitive, you now come to a rather counter-intuitive technique.

Everything That Makes the Collaborator Grow

Giving feedback on one's activities is the most important thing for a person's growth: feedback for both what he or she is doing well and what he or she is doing badly, in a timely and incisive way. The delegator becomes, in this way, a mentor who transfers his or her acquired competence over time, also giving a considerable boost to the creation of a solid relationship with the employee. In fact, the relationship between delegator and delegate becomes one in which the transfer of a real "trade" makes the delegation an effective tool for the employee's growth. In this new perspective, when I analyze the list of things that I can delegate, I will also ask myself another question: can the people to whom I am transferring one or more tasks grow through this? If the answer is yes, then I will have even more reason to continue in the direction of delegation. This will be the road that will allow both personal and corporate "take-off" because it will free potential and dormant energy, making managers, employees, and the company grow. Moreover, choosing to delegate what I am good at also forces me to raise the bar. I get rid of some tasks in which, after months or years of practice, I have perhaps reached skills saturation. By delegating, my attention turns to other things to learn. I set myself in motion and in doing so discovered another strong motive that forces me to use delegation with an even wider purpose.

All Problem-Solving Activities That Employees Find Themselves in Forehead

Another activity that should always be delegated is problem solving. Whenever an employee brings a problem to me, I don't have to solve it for him, as happens most of the time. If the manager keeps solving employees' problems, they will thank you, but they will never grow!

Therefore, through this type of delegation every time a problem arises, I will be concerned with transferring the method, resources, and objectives, rather than the solutions. Again, delegation must not be done "blindly" but must be conducted and supervised so that the employee themself finds solutions to the problem with their own resources and autonomy. Let us not replace them in the resolution. Let us only maintain driving and coaching.

Delegation as a Strategy

By choosing the activities to be assigned to employees in this way, delegation acquires a strategic function and becomes a tool to grow:

1. **as a tool to make me grow**: if I delegate something I know how to do well, I force myself to better coach others and at the same time force myself to learn something new.
2. **to make employees grow**: putting employees in front of very specific, new and different responsibilities than in the past forces them to grow because a person only grows by doing new things, with a higher rate of

responsibility and a higher rate of risk, which I accept to run together with him (and this will challenge me and challenge him, together).

3. **as a tool for company growth:** because of the two forms of growth described above.

What Not to Delegate?

Despite the greater clarity provided on what we can delegate, you will always be assailed by Hamletic doubts when approaching delegation. Do I or do I not delegate this? Do I do it or does a co-worker do it? Or even a supplier? These are very frequent questions that need precise and timely answers. We have already seen how valuable delegation can be for our own development, for those who work with us and for the company in general. In my professional life I personally clashed with a precise aspect of the delegation, when in Siemens I found myself managing about 300 people scattered all over the world: Germany, the United States, Italy, and China. At that time, I had focused on the "what" and "how" to delegate, but I had completely forgotten an important piece of information that later resurfaced and prevented me from making full use of delegation: what I absolutely should not delegate. I had no clear idea of the few priority things that I and only I had to do. Activities to be carried out personally with the utmost attention were for me in that period: care of my personal development; supervision and development of my employees; sacred time for setting up new projects and strategic initiatives; field visits and consolidation of relationships with key customers and suppliers. When I became more aware of my non-delegable priorities, I literally made a leap forward in my professional growth, because **the moment the "short list" of essential and priority activities for you becomes clear, you acquire an extremely acute vision of the things that you should do before the others**, accepting with serenity not to do or delegate everything else.

How do you know what not to delegate? You must always have in mind the three key objectives of delegation: your growth, that of your employees, and that of your company. We always remember that we only really grow when our employees grow and that our company will only grow when we both grow!

Do Not Delegate the Gold Activities

Gold Activities represent the few activities that we have carefully defined as a priority and have a high impact for our strategic development. They are the activities that provide us with a great impact on our personal and professional lives, that we are more passionate about than others, that represent real challenges, and that make us enhance our best skills, acquired or that we want to acquire to access our higher level. Often, we do not realize, however, that we waste our time on activities that have little impact, that we dislike, that we are not even that good at and that, in the end, we do not even care about as challenges. Let the few gold activities stick to you and delegate everything else.

Beware of the deception of those that are now former gold activities, those that were really Gold maybe a year or two ago but are now activities that you still know how to do well, and that may even have a modest impact, but no longer ignite your engines of passion and challenge. Well, these have become activities that you can safely delegate. Indeed, you must delegate.

Do Not Delegate Tasks to Grow Your Staff

The activities that you must do to develop your people, to assist them during a problem, to give them feedback, advice, and information, to monitor their performance should not be delegated. This means becoming a leader-coach, a leader who trains, who leads with questions rather than with orders and impositions, creating an atmosphere of mutual trust and widespread responsibility. The more people trust you, the more you will be able to delegate more and more complex tasks, because people will know that you make them grow by their side. When your people participate in training courses, stay connected and interested. Don't abandon them to the lecturer on duty. Try to personally follow their progress and the implementation of what they have learnt. Listen to them because you will get to know them better and grow up with them. Never forget the celebrations of your team's successes. Whether small or big successes, it is important to let them know that their leader can observe the good things that happen and not just what goes wrong. Become the positive energy engine of your people. You will find that they are capable of amazing things with a leader who knows how to pull the right strings in them.

Do Not Delegate the Building of Trust Relations

The best managers, professionals and entrepreneurs are characterized by their great ability to build strong relationships, internally and externally. We often observe, however, opposite attitudes. Called back by the desire to achieve results in the shortest possible time, they forget to use the number one tool to achieve them: the building of trusting relationships! With employees, with colleagues, with suppliers, with customers, with people who can add value to them and to the company. They are healthy relationships that must enter the realm of activities not to be delegated. The closing of many managers often becomes a limiting factor of their growth. Human beings tend to resemble most of the people they associate with. So, if we hang out with the usual people, and do not give ourselves the task of broadening our knowledge and relationships with other people, we will end up looking like the people we perhaps hate or dislike. Pay attention to one last detail: the healthy, solid, and strong relationships I am talking about are not developed through social networking or in a virtual way, but are developed in person, looking people in the eye, shaking hands, talking, listening, going deep. Covid-19 allowing.

Do Not Delegate Personal Regeneration Activities

Let's not forget the healthy fun! This fundamental key to personal well-being cannot be left to others. Richard Branson, iconic founder and chairman of the Virgin Group, states that "life is short and you must have fun in life to create great results. Too bad that many managers often forget this." Work environments where play disappears, where fun disappears, and where there is little trace of passion are not work environments that produce great results over time. **Playtime and psycho-physical regeneration must always be present in your agenda**. Think about what you like, what entertains you, what stimulates your passion and interest, even beyond strict business objectives. Create your own precious list of everything that comes to your mind and put something from this list every day. Every day little things, every week a little less small, every month, every year. If a culture without Sprint/Relax type principles as seen in Chapter 5 prevails in your company, you take the first step. Why is this important? Because the energy you put into your emotional reservoir in those moments will spill over into your company with colleagues, with friends, at home. And human energy is always contagious, for better or for worse. We are social animals.

Do Not Delegate Strategic and Business Development Activities

How to grow the company? What strategies to adopt? What do you want to achieve in the medium term? What do you have to do to hunt for opportunities, and the risks that your company runs? You can be the manager of your team of workers, the technical manager, or the financial or administrative manager, the production manager, or perhaps you can be an operations manager or a human resources manager, or you can be the managing director, as well as the owner or chairman of the board of directors. Whatever sphere of responsibility you hold, you should continually ask yourself this question.

What do I have to do or what can I do to develop my company or the part of the company I am responsible for?

I don't believe in ties in business. In business you either win or lose, and today the chances of winning are very high because there are so many opportunities around us, but the chances of losing are also very high. The likely competitors are very often unimaginable and just around the corner. Think about it for a moment. Look at some industrial and market contexts. You will find that those who master several industries today, ten years ago were not even on the list of companies that would challenge their respective market leaders: from hotels to restaurants, from transport to manufacturing, from high-tech sectors to services.

So, do not delegate to anyone the strategic and business development activities, the growth of your company. Stop always being only with your head in tactical activities: but every week, every month raise your head and look a little higher, with new perspectives.

Do Not Delegate Your Personal Development

Dulcis in fundo—you cannot delegate to others everything that will contribute to your personal development and training, not even to your boss or company! Even if you are in a corporate context, you should oversee your own development so that it is consistent with your vision and wishes. The path may partially or fully coincide with the one planned by your company, but you are no longer at school, where teachers were completely responsible for your curriculum. **Participate, ask, suggest, and become the protagonist of your own growth**. You choose which books to read and which training courses to do among those proposed by the company or elsewhere, outside the company. I don't want to incite revolution; I want to stimulate a sense of responsibility toward oneself and toward who one wants to become. I think the relationship between professionals and employees in a company should be one of interdependence and not dependence. Independent adults interact for a shared common purpose and together can achieve much more than they can to do alone and apart.

How to Delegate?

After defining what to delegate and what not to delegate, we analyze the main rules and guidelines to follow if we want to be sure that our delegation is effectively carried out and has a high final impact.

1. **Don't start with "what" people must do or "how" they must do it but start with "why."**
   In any field. Whether you are working with a production line operator, a salesman, a designer, or an administrator, whatever "passes" from you to him, the important thing is that you spend enough time explaining the motivations to him. The motivations that lead you to choose him as your collaborator to carry out that task, the motivations related to the importance of the task for the company, and why it must be done at a certain time and not at another time. The crucial thing is that the employee feels these motivations before going into the details of "what" he must do.

2. **Focus on the goal.**
   Before you get to the heart of do or don't do, stop, and ask yourself: what exactly is my goal? What objective does the company have in terms of content, the indicators you want to bring home, and the time required to carry out that activity. It is very important that the employee feels responsible for that objective because only in this way are intellectual, cognitive, and emotional resources activated in people who know and perceive that they are invested with a precise responsibility that you and the company are entrusting to them. That is why it is important to remember to assign responsibility for a task from beginning to end, never simple activities and isolated tasks, and with a clear measurable result. If you don't do this, it means that the bulk of the work is always up to you.

3. **Clarify how the work will be judged.**
   Sometimes you make a request, but you do not spend enough time explaining how you want it to be performed. For example: in what form do we want it? An Excel reports? A Word documents? A PowerPoint presentation? Do you want it done one way or another? Do you want it fast, albeit coarse, or is it better a little slower, but detailed and polished? You must, in essence, make explicit what will make you satisfied with the work the co-worker is doing. Depending on the area you will have to decline this third rule in a flexible and functional way, but you should always spend time on it.

4. **Give all the necessary information**.

Very often one forgets this simple common-sense rule, either because one is in a hurry, out of speed or because one is simply very superficial. Also included in this rule is the possible training that you or others will have to do to enable the best performance of the employee to whom the task has been delegated. Therefore, it is essential to ask yourself what information and training is needed for the assigned task to be performed as effectively as possible.

5. **Ask the employee to fully explain to you what he/she has understood was to be done.**

Never assume that what has been explained has been immediately understood. After having explained "why," the objective, clarified the expected quality and provided the necessary information to carry out the delegated task, asking the employee to explain what he/she understood is a way to find out, right away, if there are aspects that need to be better explained so that no time is wasted on subsequent rework or new explanations.

6. **Establish review and follow-up moments in advance with the co-worker.**

You have decided that a task must be handed in and finished within two weeks. Well, it is not smart to review at the deadline and only then discover any mistakes or small things that need to be fixed. Instead, it is good to do small periodic revisions, for example every three days, to have the opportunity to re-direct the work, to give additional adjustments and to give time and space to the co-worker to adjust the focus and do the work in the direction you want.

BONUS RULE: Feedback, Feedback, Feedback

Fast, incisive, and continuous. And no matter whether it's positive or negative feedback, it matters that it is constructive and that the person receiving them consequently feels, through your presence, the interest in collaborating to improve their performance. The result will be to have more motivated people, more connected to us by relationships of deep trust.

So, not too many personal attacks, but be careful about "what" was done and "how" it was done. Look for things that have been done well. As I have said before, it is important to surprise your co-workers with things that he or she did well, because you cannot simply make your voice heard only when something is done wrong. Learn to surprise your co-worker even for small things done well and point them out. It is important that this happens if you want functional behavior to be reinforced for the results you consider positive. It is also important to highlight also what was not done well: and then choose one thing at a time to underline, incisively, constructively; always turn it into a suggestion or proposal for correction, something that must be done to overcome the problem you have noticed. Never put too many things together if you want to keep the effectiveness of your feedback high. **We are not there to control everything; we are there to develop people**. Always remember that, because high-impact delegation is a tool for developing people and in that sense, you must use the secret weapon in your pocket called feedback. The more feedback you give, the more the relationship of trust between you and your people will grow.

The many hours spent in the company, the ever-increasing time commitment, the hyper saturation of time, the problems to be solved, the frenzy, the constant stresses, and strains of daily life, have always revealed in me the certainty of working well and profitably.

The Lean Lifestyle training has demolished and defeated these certainties of mine and the conviction emerged in me that I had become the real bottleneck of improvement.

Centralization towards me has contributed to saturating my time and reducing the quality of my work. Physiologically, my attention was absorbed by the day-to-day and diverted from the real value-added activities. I relegated co-workers and subordinates to simply carrying out assigned tasks and reporting problems. The energy around me was lacking.

When I started to apply and make use of what I learned with the principles of high impact delegation, I released the potential energy that I was suffocating and discovered different colleagues. I appeared different to my colleagues. Coaching enabled me to develop my colleagues and to be able to delegate tasks to them. I gave them confidence and was rewarded. Many of the problems that were reported to me are now solved at a lower level and in less time. My co-workers feel revalued, and the energy is high. I am now managing and organizing my time toward the important activities that are part of my role. The flywheel has started up again and is spontaneously generating improvement."

Gianluca Cunial, Ondulkart Plant Director

To Whom and When to Delegate?

But how? I delegated, I followed the rules I learned, but then in reality a very bad job was done. Is it the delegation that doesn't work or have I simply chosen the wrong person? To whom should I delegate what? And how do I choose the right person to delegate to?

Sometimes we can slavishly follow the rules described so far and despite everything get negative results due to poor "calibration" of the type of person we are dealing with when we are about to delegate something. It is very important to adopt some flexibility in the use of delegation. I cannot use the same tool in all possible situations because in everyday reality we will be faced with many people, each with their own personal and professional history. I will have to adapt my delegation methods from time to time.

If, for example, I have in front of me a co-worker with many years of experience behind, profound technical knowledge, and with whom I have a long-standing professional relationship, I will focus more on motivation, on entrusting him/her with as complete a task as possible with high levels of autonomy, on participation and on sharing operational methods.

If, on the other hand, I have in front of me an intern or a person who does not have the same experience as the previous example, a co-worker who is learning a new job, I cannot use the instrument of delegation by giving him excessive responsibilities or giving him too many explanations about the reason.

I can also do this, but I must be very careful to provide him with precise operational standards that he will necessarily have to perform and almost prescriptive activity descriptions (check lists, operating instructions, procedures, etc.). I must always keep in mind that the apprentice is learning. Similar precautions, though not so prescriptive, I will have to follow with profiles that have evolved but have never done the type of work I am delegating.

In Depth

According to Eastern martial arts, every moment of human learning and evolution can be described in the set of three phases called *SHU-HA-RI*. The first of these three phases—SHU—focuses on breaking down into small, simple elements of something bigger to learn. Think of the learning phases in music, in an artistic or sporting discipline, where it is considered normal to start with small, clearly decoded and standardized elements that must be repeated over time, under the close supervision of the "master"—coach—before proceeding to the HA phase in which the learner proceeds to assemble the various basic elements, again with the slightly looser supervision of his teacher. The third phase, the RI, is where the apprentice can be free to venture out with more autonomy and perhaps even creatively reinterpret some steps. This third stage, however, is very time-shifted and it is the teacher who decides whether the pupil is ready or not. Therefore, in the case of someone who is faced with complex work that has never been done before, I must take great care in selecting the simplest elements to be executed precisely. Instead, we often make the gross mistake of using the same delegation methods for all people, because delegation is not conceived as a tool for learning and developing new skills.

Having clarified this aspect, let us now try to better interpret the various possibilities we have in identifying the recipients of our delegation. First, I would like to make it clear that it is not important to have people in close organizational hierarchies at all costs or to have many staff at our disposal. What matters is being able to refer to the people with whom we develop—or could develop—our technical and managerial relationships. Perhaps you are a technical manager, with few or no staff in your organization, but you still have people to delegate to: suppliers, service providers, colleagues and members of the various project teams in which you are involved. Let's get used to enhancing both internal and external collaborators. And let us remember to consider the company as a link in an extended chain. Often outside our usual boundaries there are fast and efficient professionals hiding.

For example, ask yourself which colleagues you can entrust with activities that go beyond your sphere of competence, perhaps in a project area where you are working cross-functionally with several functions involved in a common project. It is **not necessarily the case that the people to whom you can or must delegate fall within a direct organizational sphere of supervision, but there are many possible indirect spheres of supervision: colleagues, suppliers, bosses, partners, consultants, customers, etc**.

If you fall in love with high-impact delegation as a lever for personal and organizational development, which I very much hope you do, you will find that you have an army of people ready to help you if you only decide to value them. Inside and outside the company. When to delegate? As soon as possible! To immediately trigger the growth mechanism of yourself and your company. The "muscle" of the delegation must be trained more and more by those who want to move decisively in the direction of Lean Lifestyle, to get more performance and better quality of life.

The Reverse Delegation

You can delegate something to your boss—attention bosses in reading!—because in some cases what is called "reverse delegation" may be practiced, a thin field of demarcation between delegating not downwards or sideways, but upwards. In this case, it is the employee who delegates to his or her boss.

What can the employee delegate to his boss? He can delegate answers to be obtained in a certain precise time, he can delegate priorities to be indicated so that he can direct his work in the best possible way, he can delegate satisfaction criteria to be expressed so that he can do the work to best meet his expectations, he can also delegate monitoring and follow-up activities. I remember an anecdote I experienced during a project in a client company a few years ago.

An employee had applied the principle of slotting, which we saw in depth in Chapter 5. He had learnt to divide his working day into portions of time with high concentration and high focus, with great results in terms of quality of work and time saved. However, he often found himself "trapped" by his boss who intervened when he least expected it by demanding immediate attention. Here is a statement he made:

> I tried to arrange the agenda as best I could, but my boss comes and breaks the plans I made. And so, there is no agenda that I can keep because you must answer when he calls, you have to immediately go to his office when he calls roll call, you have to answer the phone when it rings and it's him, you have to interrupt everything to meet his demands suddenly.

In this case we used reverse delegation. How? We had the employee propose, almost without being too conspicuous, to his or her boss's meetings on specific dates and times: "can we meet today at 5 p.m.? Tomorrow at 6pm?" We started setting meetings where it was the collaborator in advance who asked for an audience, slowly turning these hearings into recurrences, and thus setting portions of time in which the boss was "forced" to meet the collaborator who arrived at these meetings already with a list of questions and already with a list of topics to be discussed. In this way, even the boss ended up concentrating, after a few meetings, all the questions to be asked and the answers to be obtained in that fraction of time. The enormous variability that previously existed has been greatly reduced. So, in this case, it was not the boss who asked for specific meetings, but exactly the opposite happened, the co-worker using the concept of reverse delegation in a particular way, forced the boss to give him an audience and make the relationship flow in a more efficient way.

Why Should We Delegate Hard and Well?

Very often the problems we have with others, whether in managing delegation or in the difficulty of letting go to a massive delegation, reflect possible problems we have with ourselves: trust, self-esteem, awareness. Through the exercise of delegation, we will learn to deal with it and develop personal skills that are fundamental for our growth.

Imagine you are a father struggling with a child who must learn how to ride a bike. It is not only the child who has to learn to ride a bike, but also the father who must teach. The same thing will happen in the delegation and the same circles of learning and self-esteem growth will be triggered. Both in the one who teaches and in the one who learns. It may happen that the child who is learning to ride a bike falls. In that case the father teacher will be at a crossroads: he may become impatient, scold the child, perhaps shouting at him that he has not yet learned; Or he may look at the child on the ground and congratulate him for having managed to ride a few more meters than the previous time, and simply help and encourage him to get back on the bike and continue. Only in the latter case will the father's behavior lead to the creation of virtuous chain circles. You will often be faced with similar crossroads in the management of your employees. The choice is yours. Through the quality of your feedback, you can generate virtuous circles or create silent vicious circles that slow down and hinder mutual development.

Many people are so busy doing all the jobs that come their way, they feel so full of responsibility for all the "burden" of tasks they bear, that they convince themselves that they do not even have time to stop and explain the job to someone else. They want, in their hearts, to get it done as quickly as possible, and they think they can do it by going it alone. On other occasions they delegate the work to others, but do so in a hasty manner, assigning the title of the task, the deadline and often without even clarifying the expected result. This way of acting almost always leads to mistakes, rework and wasted time following delegation. This is not delegating. It is abdicating one's role as a manager, and certainly not acting as a Lean Lifestyle Leader.

I believe that there is always enough time to delegate well. From now on, stop saying that you don't have time to delegate a task clearly. Remember that knowing how to carve out enough time to delegate well is the most effective use you can make of your time, if you want to achieve more energy for yourself, more results and better quality, while at the same time growing yourself and others.

One last consideration at the end of this chapter: there are always side options. Where there is a process or activity that needs to be carried out, there will always be someone who can help you in its implementation, partially or completely. But it is you who must seek out and explore your virtuous connections inside and outside the company. And in the end, you will also find that the activities that are looming over you today and weighing down your days can be carried out much more effectively and efficiently overall.

Lean Lifestyle Story: Orogel

Delegation That Makes People and the Company Grow

The transition from Lean application in production processes to Lean Lifestyle to apply Lean Thinking in and for people, the transition was a natural one, given the extremely welfare-conscious environment within the company. The Lean Transformation project at Orogel, the Italian market leader in frozen products, started in the jam department in 2016, later expanding to other departments. From the beginning, the challenge was to accelerate the execution of improvement ideas, develop the skills and versatility of the workers, to grow people, at all levels, through delegation and the construction of structured organizational routines.

> "When I think of the project we have carried out," recalls **Andrea Maldini**, Coordinator of the Orogel "Smart 360" Confettura System, "I often compare it to the plowing of the fields: this is because, thanks to this simple activity, without adding too many fertilizers, the soil becomes more fertile, ready for the new season. So, in Orogel we started with what we had to record important improvements in terms of productivity and working well-being."

The improvements included the sharing of "tricks of the trade" by department heads with operators; the creation of checklists of optimal key behaviors; the implementation of a system of observations and feedback on the field and the start of organizational routines of weekly alignment. In particular, the latter enabled the heads of the jam and packaging departments to develop their respective department heads.

For example, every Monday the managers prepare the meeting by giving precise indications to the department heads, who thus arrive at the meeting, two days later, having clarity of the projects under discussion and of the numbers to be prepared to support them and at the same time on the solutions identified with respect to the problems detected.

The organizational routine has made it possible to reduce the time required for meetings, but above all it has consolidated the habit of department heads to be pro-active about problems. The meeting focuses, in fact, only on the solutions proposed by the department heads. Decisions are made through structured discussion, formalization of ideas, analysis, and definition of shared action plans. No time is wasted in analyzing the problem because this is already done in advance. More responsibility, more focus, and therefore more results in terms of production efficiency. Another acquired benefit: no more returning to problems and thus no unnecessary energy wasted. The quality of managers' work has improved because they are no longer the bottleneck of problem-solving actions and at the same time department heads have seen the importance and value of their role recognized.

The process of delegation to the line operators has helped to increase the plants' efficiency. For example, the following performances were recorded in the packaging department:

- The OEE improved by 14%.
- Speed losses have been reduced by 52 %.
- Failure stops have been reduced by 18%.

The complete versatility of operators has been achieved with their full satisfaction and that of their tutors. In the packaging department, in 2020, in 16 weeks, the overall target for acquired skills was reached (Figure 6.1). Finally, motivation and willingness to learn have increased. Significant testimony collected at the end of the project:

> We learned to communicate and confront each other. We are co-responsible for production results because we now work with different criteria and with a method that has eliminated wasted time. Before, we just did things, now we "understand" how the machine works and we are more autonomous. I'm glad that my

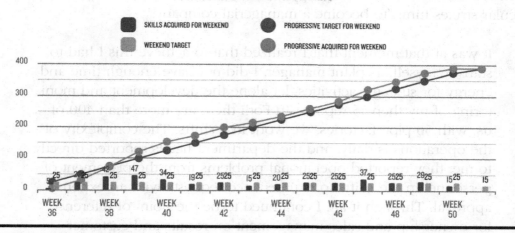

Figure 6.1   Skills acquired by machinists in the packaging department.

initial anxiety was dampened by a tutor who had the right manners and tones, the patience to explain, and gratified me. Sometimes it is said that we should not wait for the consent of others, that you just have to be aware of what you are doing: I instead believe that this is only partially true, I believe that in a team work there is a give and take between each of us and feeling gratified is the only thing that can give you the certainty that you are doing a good job, otherwise you will remain insecure. With the fear of making mistakes, you always take a step backwards. With the encouragement to do more and more independently, on the other hand, you stimulate personal improvement and growth.

"With the Lean Lifestyle methodology, we became aware that what we had been doing for years could be done in a different way, in a better way," concludes **Michele Pagliarani**, Orogel Jam Production Manager. "In fact, one of the hardest things to change is habits, and so we have acted in people's behavior to improve our way of working and thus production performance."

Lean Lifestyle Story: Marcegaglia

More Autonomy to Go Faster

In 2019, **Lorenzo Bonacina**, plant manager of the Marcegaglia tubing plants in Forlì and member of the board of Marcegaglia Specialties, has just defined the three-year strategic plan and among the objectives one in particular strikes him: "to become a managerial company."

> It was at that moment that I realized that to achieve this I had to start by myself, as plant manager. I did not have enough time and energy for strategic activities, let alone the development and monitoring of my then 15 reports. In Forlì there are more than 400 of us, with 50 pipe factories, we work in 3 shifts. The complexity of the operations is daily, and the department heads reported directly to me: they reported operational problems from the placement of personnel in the shifts to maintenance requests with an expense approval. Through it all, I continued to be the point of reference for ownership and sales management on results and customer promise respectively. I was literally drowning. To change this situation, I had to start delegating in a structured way.

The initial analysis was clear and precise: the director had become a bottleneck for making certain decisions; not all the employees had clear responsibilities and therefore felt little involved and motivated; some employees were undervalued compared to their potential. New ways of working were needed:

> I arranged weekly meeting routines with each of my then 15 reports so that on the one hand I would give them quality time to involve myself in their topics and on the other I would give myself the privilege of not being interrupted every five minutes. In these meetings, I began to be joined by those I thought would be my future Production and Supply Chain Managers.

The next steps to follow up the structured delegation process were:

1. to map the managerial competencies of all potential new leaders;
2. to cascade all the indicators that would then characterize and measure the responsibilities of the entire new chain of command;
3. to write the new organizational chart and all the new job descriptions;
4. to meet with the company's ownership to endorse the work in progress;
5. to communicate the change to each employee.

The plant manager emphasis:

> For us, and in particular for me, it represented a significant change of pace: I went from 15 to 7 employees. And today decisions are faster and closer to where things happen, also thanks to the new leaders. I can take better care of the advancement of the company's strategic plan and give the help my people deserve, and—only rarely—do I have to accelerate when the second lines do not respond as they should. I don't have the proof to prove it, but I honestly don't know how I would have coped with the pandemic in the factory if I hadn't had my head cleared of the brisker operations and employees capable of making decisions on their own.

Lean Lifestyle Story: Lucchini RS

Delegate to "Fly" Themselves and Others

Alberto Ronchi is the design manager of the Railway Division at Lucchini RS. His Lean Lifestyle journey has centered around the issue of delegation, and it was already clear from his comments in the classroom during the initial training that this would be his challenge to win. And he needed it, not only to gain time and energy to devote to activities that were truly distinctive for his role, but also to regain a free-time-work balance that had been severely jeopardized by the "e-mail addiction" and "absurd hours" that he had been doing for years.

> "The reflection on gold activity was the first stumbling block," he recalls. "I had considered many operational design activities as gold because that is the heart of our function, but then reflecting I realized that my role was another and I only chose those really necessary. I also realized that I wasn't devoting time to some important things such as growing employees: stopwatch in the hand, I wasn't even getting to one hour a week! I immediately started to get active on delegation even though I felt a bit lost in this role. By experimenting a bit, I found my method, and above all I tried to find a way to shift the workload in a sensible way. Little by little it started to work well."

E-mail management was the second stumbling block.

> I was overloaded with work; I went into all the operational com-
> munications and became the single point of contact for any ques-
> tion: everyone asked only me. Consequently, I was overwhelmed by
> e-mails: I spent almost 5 hours a day on them and of course I was
> an unrestrained "multitasker" because wherever I was, I was trying to
> "get ahead." The time spent on e-mail has now decreased so much,
> almost by 40%; I really wasn't hoping for that at the beginning, and
> now I am working on it to pursue my target of halving it than before.
> I started with slotting and trying to avoid multitasking, although,
> I confess, this remains a weakness for me to improve. But one of the
> keys to e-mail has been delegation. For example, a lot of customer
> communications that I used to do in bidding and planning are now
> done by my co-workers. In the simplest cases they operate autono-
> mously, and we have agreed that they don't need to copy me; if the
> situation is more complex, we can discuss the approach verbally and
> then they do it. My head clicked when I started to put myself in their
> shoes, thinking about how they could be perceived by colleagues in
> the company and by customers. Before, I only saw the perspective
> of efficiency: if I had intervened, I would surely have done it sooner
> than to explain it to them, also because, I confess, "then maybe they
> wouldn't have done it as well as I did." It was an enlightenment;
> I was really clipping their wings. I had to get out of the logic of
> "I don't delegate" because I am better at it. It's true what I learnt with
> the Lean Lifestyle: e-mail in cc is often a problem of delegation.

Thanks to the work done on e-mail and delegation, along with several other Lean Lifestyle strategies, in just four months Alberto was able to
reduce his workload by almost three hours a day (without impacting on the
size of his department) and was able to carve out time for gold activity that
he did not do before and for his private life.

> "I feel it's much better now, I can see the effect of the fewer hours,
> but above all I am less stressed and manage my time better," he
> concludes. The opinion "stolen" from one of his staff members is
> a fitting seal of his growth as a manager and Lean Lifestyle leader:
> How is Alberto doing with delegation? Wow, he delegates a lot
> now!!! And we are learning a lot.

Chapter 7

The Simplifying, Low Number of Interruption Company

What can a company do to help people successfully implement the principles of simplification and agility behind Lean Lifestyle? If we want the principles seen so far to become an integral part of a new way of working, it is necessary to take organizational steps with a holistic approach.

First, it is important to turn the spotlight on the entrepreneurial goal and share it within the company: to simplify, streamline office processes and reduce interruptions in a systematic way to foster real well-being and productivity for individuals and the entire organization. By sharing this ambitious goal, many actions can be taken at the organizational level to facilitate the implementation of the individual processes and routines seen so far in the book.

This first step corresponds to the breaking of an obsolete and harmful paradigm that still survives in many companies and in the beliefs of entrepreneurs and managers: **the more we work—and the more we make work—the more tired we get, the more results we get. Nothing could be more wrong**.

The belief we should all strive to make our own is that the better we work, the more results we achieve.

We often work very badly. In all companies, the perceived increase in workload hardly ever corresponds to an actual increase in output. This does not mean that people do not get tired, quite the contrary. It means, instead, that they often get more tried to produce less! According to the studies of Gerald Weinberg,[1] referred to as the computer psychologist, huge amounts of time and energy are lost with each project change due to constant "context switching." These losses, as seen in Figure 7.1, can

DOI: 10.4324/9781003474852-7

NUMBER OF PROJECTS MANAGED SIMULTANEOUSLY	PERCENTAGE OF AVAILABLE TIME PER SINGLE PROJECT	LOSS DUE TO CHANGE FROM ONE CONTEXT TO ANOTHER
01	100%	0%
02	40%	20%
03	20%	40%
04	10%	60%
05	5%	75%

Figure 7.1 Loss of time and energy due to "context transitions" according to the studies of Gerald Weinberg.

be as high as 75% of the overall time available. And it is not necessarily necessary to change projects to suffer these losses but simply to have a workflow, even a mono project, that is highly fragmented and frequently interrupted by e-mails, telephone calls, requests, distractions, and meetings.

These losses of human time and energy are very often overwhelmed and undetected by any objective assessment, and the relative organizational countermeasures are rare.

We struggle to chase a huge number of departing trains every day, tiring ourselves much more than we should, often accumulating delays and sometimes missing arrivals at our destination.

To address and improve this situation, new individual awareness, appropriate measurement systems, and, above all, organizational countermeasures designed at company level are needed. However, none of the identified actions will be effective or will really be experienced as indispensable if the ownership and managerial leadership of the company does not carefully reflect the underlying motivations and problems that you want to solve. For the diffusion of agile working principles, the first obstacles to be removed are the mental ones. And these obstacles must first be removed at the top of the company. In the following pages, we will look at six organizational proposals, born from the experience we have had over more than 25 years with many companies grappling with the attempt to spread agility and well-being inside:

1. agility and well-being metrics;
2. the removal of environmental obstacles and the creation of spaces for agile work;

3. training and continuing education on the subject;
4. definition of company routines and compliance with availability windows;
5. knowledge management;
6. checklists and FAQ.

Agility Metrics and Well-Being of the Company

In manufacturing environments, there is a long tradition of performance monitoring. The effectiveness and efficiency of production is measured in various forms and, although improvable, they are often sufficient to trigger process control and improvement processes. Companies have data on how much has been produced, shipped, or put into stock, the number of rejects, how much has been scrapped or reworked, and how efficient the production staff have been against assigned targets and how much time has been lost in product changes on the lines.

In manufacturing companies applying Lean Manufacturing principles, the first cultural element that is transferred is that you only improve what you measure. Therefore, it is relatively easy to talk about output quality, productivity, and efficiency thanks to the introduction of KPIs—key performance indicators -: from the measurement of overall plant efficiency—OEE Overall Equipment Effectiveness—to the timely detection of single workstation or individual operator defect rate.

And in the endless world of intellectual work, of so-called indirect work and of offices, what is being done to effectively assess real productivity? And what do you do to measure people's well-being?

These two questions are becoming more and more important in recent years, and will become even more so in view of the fact that the amount of real work is increasingly shifting toward the indirect slice of work, so much so that the classic distinction between direct and indirect work, typical of the 1920s when the Taylorist theories of fragmentation and work organization were born, based on a model of mass production that is no longer relevant today, also appears obsolete. In those years, most workers were direct, that is directly linked to the production carried out, and thus a variable cost of production. A minority of workers, blue-collar, white-collar, and managerial workers, were dedicated to the maintenance, management and development of the entire production, infrastructure, and commercial system, representing the so-called fixed costs of production. So much so that it was not far-fetched to pass these fixed costs over the large amount of variable production costs to calculate the cost of products. Today, the opposite happened. Fixed costs and related investments have become the bulk of the cost, and this trend is on the increase, even though many companies still insist on calculating the cost of products as they did a hundred years ago. Technologies, automation, speed of development and market launch projects, fragment for new ted and small batch production, extreme complexity, and variability of the environment, increasing need for management and maintenance to keep the system efficient and constantly improving require an ever-increasing amount of indirect work.

It is people who move work forward, not processes or tools.

If we want to measure at the same time how effective we are in the work of the entire company, how much the results and the well-being of the entire organization are growing over time, we are forced to introduce metrics that are still considered unusual, but indispensable. If we want to apply also in the world of intellectual work the concept that we only improve what we measure, we must get out of some limiting beliefs that bind us to patterns of evaluation that are not functional to the new context in which we live.

Examples of Indicators for Measuring Productivity and Well-Being

Let us look at some examples of indicators that can be introduced to measure productivity and well-being in the modern workplace.

Durations in the evasion of the company's transversal key outputs

Examples:

- response times to customer quotation requests;
- Total development time for new products;
- innovative concept generation times;
- market launch times;
- Project closure times across different business functions;
- time frame for the annual budget.

It is important to identify outputs that have a direct link to the business impact, but that at the same time require the participation of different business functions. Only in this way will it be possible to monitor the performance of teamwork and the real ability to quickly produce results in complex contexts. **Measuring times and durations is crucial if we want to shift people's focus on results and agile ways of achieving them.**

Dealing with problems, making decisions, executing established actions, verifying their effectiveness, and repeating this cycle over and over again are at the heart of modern work. We need to measure how quick and productive we are in this type of actions.

Closing capacity of ongoing activities

Examples:

- number of projects behind schedule;
- average latency of job evasion;
- timeframe for implementation of improvement proposals;
- percentage of shares not closed compared to the total.

These indicators represent a litmus test of the way people work in the office. Faced with delays and poor performance, the most common alibi that one sees being erected in all companies relates to the excessive amount of work and activities in progress. We are all ready to complain about how overburdened we are. That may be true. But after all, who knows what is the "right" level of workload when it comes to human labor? And above all, who can say precisely how much of the load comes from wrong ways of working or wrong processes? Let us remember that the conditions of rapid and frequent change of subject matter cause an increase in the perceived workload that unfortunately even results in chronic stress conditions in most cases. In highly multitasking contexts, productivity losses can be as high as 50% of the net time required. Therefore, since it is difficult to measure how many distractions and how much multitasking is perpetuated damagingly and continuously in all work environments, it is highly advisable to at least measure the effects on performance of the way we work. By doing so, we will shift the focus to improvement and realize in practice that we can reduce our workload by feeling engaged in only one task at a time and forgetting everything else for those hours when we are busy closing that task. The more things we see closed, the more we will want to close and the more we will convince ourselves that we can work better.

Parameters of effectiveness and efficiency of meetings

Examples:

- total time spent in meeting;
- persons present on summons;
- the extent of delays in starting and ending meetings;

- measurement of finished meetings with unmarked agenda and action plan implemented;
- measurement of the off-topic topics covered.

The sense of this type of measure is trivial: we spend most of our time in long and exhausting meetings, but generally we do not measure in any way the quality of the time spent, nor the effectiveness and efficiency of the meetings themselves. It is like having a labor-intensive production plant, but not measuring productivity and performance in any way.

In all the contexts in which one has started to monitor certain parameters of meeting effectiveness, the result has been an increased awareness and consequent improvement of these parameters over time due to actions and countermeasures taken without even too much effort. You only improve what you measure.

Parameters of effectiveness and efficiency in e-mail management

Examples:

- number of global e-mails circulating;
- number of e-mails circulating inside and outside working hours;
- total space occupied in the network infrastructure.

These indicators are useful to understand that, generally, an increase in the number of global e-mails circulating in a company does not correlate with an increase in turnover of the same. There are companies whose turnover is stable or decreasing, but where the number of circulating e-mails strangely increases, even though there is no increase in the number of people employed in the company. There are companies whose turnover rises over the years, but not as steeply as the overall volume of e-mails and the expenditure for the operation and maintenance of the necessary network infrastructure. Why does this happen? We work more and more, but worse and worse. People copying unnecessarily, multiple e-mails on the same subject, poorly written and therefore needing corrections and additions, or messages that could be replaced more effectively by a phone call. And I could go on and on. Certainly, guidelines and countermeasures for the agile use of e-mails are necessary, but only by measuring will we be certain of the correct focus of people.

Amount of hours worked

Example:

- quantity of ordinary and extraordinary hours worked, in absolute values or as a proportion of turnover, subdivided for manager and not.

Probably, in the not-too-distant future, working hours will be abolished both as a concept and as a metric for evaluating people's performance. Think of what is happening with the spread of Smart Working, which has accelerated violently during the Covid-19 pandemic period. We are strangely still tied to the paradigm of the 8-hour working day, which then becomes 10–12 for different managerial brackets. **And we feel guilty when we work even one hour less, instead of being happy**.

It is as if we wanted to evaluate an athlete on how many hours he or she trains and not according to the chronometric result or the amount of weight lifted. A person or a group of people should be rewarded every time they show improvement in their performance—sales, bids, projects closed, problems solved—with less time spent. Companies would take off if this happened. Results and well-being would start to increase simultaneously. Finally.

Parameters for the dissemination of delegation

Examples:

- number of actions and projects with responsibilities uniquely assigned to non-apical profiles;
- number of ATRED carried out (is a tool that we will analyze in Chapter 10) and discussed with the respective managers;
- amount of intra- and inter-function alignment meetings;
- Amount of improvement ideas generated;
- number of potential replaceable managers for each function.

The measure of the diffusion of the delegation is rarely adopted in the companies in structured way, but it is one of those that can radically change the very concept of managerialism. A client and friend of mine, Beppe Scotti, an entrepreneur in the food industry and a long-time restaurateur, transformed his PC screen saver with the words: "I work to make myself useless." This resolution represented his change of mindset and has done him good. In about 10 years, from 2009 to 2019, he went from 4 to 11 establishments, starting the first Lean Restaurant in Italy and supplementing his company with new business lines, including a roastery, a brewery, two farmhouses and other lines of production and marketing of organic products and beverages.

Parameters of psychophysical and emotional well-being

Examples:

- assessment and survey results administered to large population groups corporate

■ results of specific assessment of the degree of fitness and health offered by the company on a voluntary basis.

These are periodic, ad hoc check-ups that detect objective and subjective elements of people's individual well-being. Various assessments can be carried out that provide answers on the state of health, the degree of fitness, the level of stress, the ability to concentrate, the emotional and mental well-being of people, their degree of risk of burnout. Why deal with these elements in the company? Many studies show that there is a strong correlation between the degree of well-being of people and productivity in a company. In practice, the better one feels the more productive one becomes and the more absenteeism and the percentage of work that is paid but has little impact on results is reduced.

Level of people involvement and organizational well-being

Examples:

■ degree of sharing of objectives;
■ measure of the level of participation and involvement in company initiatives;
■ level of staff motivation and sense of belonging;
■ level of job satisfaction;
■ measurement of the level of mutual trust;
■ exhaustiveness of company information made available;
■ level of perceived fairness and meritocracy.

A company can be almost a second home for employees who feel it as their own, where people who are satisfied to get up every morning and sit at their desk or workstation, become able to withstand changes in the external environment and adapt, also increasing productivity. A good organizational climate, in fact, encourages concentration on the tasks to be carried out and on personal relationships, even outside the workplace.

How to move between all these possible indicators? Adopt them all? No, no need. The most sensible way is to make an initial macro-assessment that includes both performance and well-being aspects, understand the situation and the initial data collected and start monitoring areas linked to one or more dedicated improvement projects. Change, both in people and organizations, is fostered by small triggers that cause cascading effects. We must not demand it suddenly and in drastic ways that might even create rejection effects.

Lean Lifestyle Story: Labomar

Measuring Agility and Well-Being in the Company

Labomar is a company based in Istrana that for 20 years, under the leadership of its founder Walter Bertin, has offered its customers solutions in the field of food supplements, medical devices, cosmetics, and food for special medical purposes. Starting in 2017, Labomar decided to evolve its way of working, not only by adopting Lean management tools and practices, but also by gearing the entire organization toward a Lean Lifestyle Company transformation.

Among the activities carried out in 2020, the Lean Lifestyle Score Card project was particularly important: a system for measuring agility and well-being in the company (Figure 7.2).

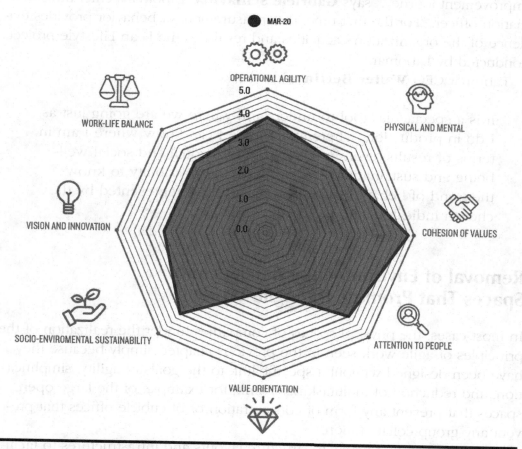

Figure 7.2   Radar chart of the Labomar Lean Lifestyle Scorecard.

In the light of the results of assessments and subsequent actions in terms of Lean Lifestyle projects conducted from 2017 to 2020, it was in fact decided to build a system of analysis of the behavior and results of people in the company in order to understand the company's culture in a precise way, thus overcoming the limitations of the questionnaires that basically investigate people's opinions.

The first step was to choose "what to measure" in terms of behaviors and results. Eight main dimensions were then identified, including operational agility and value orientation. Twelve people were identified to be operationally engaged in data collection and measurement.

The dashboard created allows the company month after month to collect the chosen indicators and take consequent actions to improve results. A clear and precise responsibility was also given to Lean Agents to contribute to the improvement of the different dimensions identified to support change.

"The introduction of the Lean Lifestyle Score Card was a keystone to our improvement journey," says **Gabriele Schiavina**, Labomar's chief transformation officer. "For the first time, the measurement of behavior provides evidence of the organization's activities and results in the Lean Lifestyle projects conducted by Labomar.

Labomar CEO **Walter Bertin** adds:

> It is a spectacular tool to understand where we are going just as I do in production, where at any time I can know where I am in terms of results, even in the area of individual and social well-being and sustainability issues I have the possibility to know the trend of behaviors target of observation represented by the chosen indicators.

Removal of Environmental Barriers and Spaces That Promote the Agile Work

In most cases, the physical spaces of companies hinder the realization of the principles of agile work seen in the previous chapter, simply because they have been designed without a specific link to the goals of agility, simplification, and reduction of multitasking. Think, for example, of the large open spaces that prevent any form of concentration or of cubicle offices that prevent any group collaboration.

The company that wants to structure layouts and infrastructures to facilitate agile, high-concentration work, to encourage the minimum number of interruptions and the maximum well-being of people, must always ask itself two key questions.

- What can promote productivity and well-being in our working spaces?
- What can hinder productivity and well-being in our working spaces?

With a good dose of creativity and unconventionality, we can equip ourselves with colorful and well-soundproofed isolation headphones, to signal that we are in sacred time mode and dive into a high concentration sprint. We could design an optimized layout to encourage both communication and concentration, creating dedicated spaces for individual and team sacred times, spaces for relaxation or for motor activities, for reading, for socializing and for isolation. The very position of people should be assessed according to their actual movements and relational needs with other people in the organization. It is pointless to put people close together who will talk to each other once or twice a week and people who must talk k to each other every day.

It is no longer rare to find so-called focus rooms in offices to ensure the availability of single or small spaces designed to promote concentration, soundproofed, and usable by everyone on a rotating basis.

These are not traditional meeting rooms but ad hoc spaces for working in "sacred time" mode that can also be used to convey other messages, such as in this space no phones, no multitasking, and no distractions. Other companies set up equal sacred time slots for everyone in the office, for example, from 1.00 p.m. to 2.30 p.m. In this time slot, no meetings, no interrupting colleagues, and co-workers. Everyone is free to concentrate on his or her own task that is considered important and a priority for that day. In all of this, an advice: never leave the architects or designers on duty to work alone, but supplement their work with the functional analysis linked to the agile working processes you want to promote in your company.

Lean Lifestyle Story: Timenet

New Business Spaces as a Collective Expression

Timenet was founded in 1996 and today is a telephony operator for companies. An organization of over 50 people and a turnover of almost 10 million euros.

Franco Iorio, president and founder, chose to bring the Lean Lifestyle into the company at the end of 2018 with a project that involved the entire organization.

> I took the opportunity of the project to unify two local offices in a single space, to introduce the Lean Lifestyle methodology. My aim was to simplify and streamline daily work activities, increase motivation, and simultaneously improve people's well-being. I wanted a location that would be the best place to spend one's days, like your living room.

The application of the Lean Lifestyle method consisted in mapping processes with a focus on critical issues related to space (foreseeing the company's development over the next three years with an accordion logic that allows spaces to be rethought based on the actual number of future employees) and relational dynamics. The main actions of the project:

■ **Envisioning layout**: the values experienced by people in the company and the opportunities and obstacles to live them in the existing workspaces have been identified.

■ **SWOT analysis**: the strengths and weaknesses of current spaces, as well as the opportunities and risks of future offices through an extended engagement process.

■ **Mapping relationships**: the intensity, frequency, and importance of proximity between offices have been mapped to identify the best layout that expresses people's maximum potential.

■ **Sizing of offices based on expected growth** in the coming years, in line with the Timenet business plan.

■ **Creative sessions and expression of people's energy**: workshops in which maximum freedom was given to the creative energy of individuals, engaged in the characterization of spaces oriented toward well-being and excellence, were able to give life to their creativity. In practice, business processes have been associated with the places where they occur, exploring their criticalities to avoid wasteful use of space.

The project has led the entire organization to be the protagonist of the site, an active voice in providing the indications then translated into architectural and functional requirements.

In the various sessions, a clear indication emerged from the employees to have small rooms for high concentration activities, niches where there is acoustic insulation and an environment for physical activity and relaxation. Banned open space, but spaces with few people. The entire organization was and felt itself to be the protagonist of the overall "design" of the future workplace from a new point of view: space not as a simple place of work but of life, of the expression of one's "self," of one's "relationships," in which performance is doubly linked to individual and organizational well-being.

Training and Continuous Education on the Subject

In this context, unfortunately, macroscopic errors are very common in most companies. Have you ever seen an athlete improve through just one training session or so? Or have you ever seen an athlete who trains for a few years and then stops training? You will hardly see him or her go into competition as if nothing had happened. In the professional world, on the other hand, people stop training and studying assiduously once they have finished their school and university studies. They may do some training in later years, but in many cases, there is no connection with immediate job application. It is as if an athlete learns a new technique during a workshop with a new coach, but then goes back to training or competing and does not apply and experience anything he has learnt. What would be the result? Surely the new technique will be forgotten in a short time. Over the years, I have had thousands of people acquire many of the concepts you are reading about in these pages, but in the vast majority of cases I have observed a common phenomenon: only when people set out to cultivate and "train" the notions they have learnt, is it possible to observe consolidation of new real skills and thus observe behavior that changes permanently. **It is therefore not enough to do a course or read a book to enable lasting behavioral changes. It is necessary to design a real gymnasium** in which one continually experiments with new notions, receives feedback, adjusts one's focus, repeats concepts, applies them again, studies new elements and so on. This cognitive-behavioral gym should, in my opinion, become a constant in everyone's life just like any sport or simple new movement that we want to be part of our routines. The key step in this virtuous cycle of continuous training and creation of new skills to enable new behaviors is to create a structure of internal and external mentors to support and accelerate learning over time. These mentors will become the "enzymes" necessary for the successful digestion of new principles.

Lean Lifestyle Story: Lamec

The Strength of a United Learning Team

"We are turning my idea of a company where people are not a number into concrete, structured actions. The Lean Lifestyle approach has enabled me to make this purpose tangible. How? Focusing not only on processes, but also on the well-being of people and the value they can generate. And when I think back to the year 2020, I realize that the serenity and the spirit of cooperation achieved in a context that was certainly not easy, was an achievement to be proud of. Involvement and sharing, incorporated as company routines in project management, have allowed us not to lose our bearings, never to stop."

In the 4,000 square meter factory in Cimavilla di Codognè, **Yuri Perin**, owner of Lamec, a company specialized in mechanical processing for the catering sectors professional and naval, he does not simply see a company, but a team of people to lead and train to work together.

"Of course, we had tried that in the past, but it didn't work because we couldn't keep our promise and therefore paid for the mistrust generated in people," he admits. "Today, however, we monitor the actions, measure the activities and examine the proposals that come from each of us. Everyone contributes and feels useful. The periodic meetings of alignment and sharing of projects/company progress, organized, and prepared in a standardized manner, have created such an energy that even those most anchored to 'old-fashioned' ways of working have been transformed from passive to active players. With regular meetings, organized and prepared in a standardized manner, the change was consolidated."

The growth of employees' skills has become a strategic asset and pursued in a structured way and this "has generated well-being in the involved people and economic results in terms of increased productivity and reduced costs."

Among the activities carried out in the offices was training on the main Lean Lifestyle tools:

- ATRED and consequent sacred time for higher impact activities;
- Weekly slotting/scheduling for e-mail and project management to avoid rework;

- Defining role responsibilities and synchronizing activities between colleagues to reduce waiting times;
- Elimination of dead time due to tool time (example: start of the day with the launch of the consumption control and materials reordering software, which consolidates during the warm-up of other activities without having to have the PC locked).

"More smiles and serenity. More well-being at work," Perin concludes. "When did I notice that something had changed? When I realized that people started to work as a team also outside the work context, when I read in the messages of our internal chat the pride and happiness of working together and with ambition, all together."

Definition of Business Routines and Compliance with Availability Windows

When you must do any errands, at the post office or supermarket, the first thing you do is to check the opening and closing times of your destination. And of course, you don't leave the house if it is closed. In business—and when we are in a "work trance"—not everything that is obvious is considered and executed. On the contrary. If we think about asking something to Tom, we just do it. If we want to meet Harry to discuss that unfinished business, we go to Harry and do it. And on the other hand, it will be very likely that Tom will answer us, and that Harry will give us his instant willingness to talk to us. I don't want to push you to stiffen and formalize all kinds of meetings and communication exchanges in the company, which is impossible even if you wanted to. I'm just trying to make you reflect on the organizational advantages of consolidating a very simple but effective behavioral habit: getting used to openly declaring and respecting each other's availability windows. I am available on Mondays from this time to this time, and when are you available? I have these windows of availability, and you? Often, it is taken for granted that everyone is always available, and not only in the company. While shared electronic diaries, from Outlook to Google, allow you to set limits, few do. Having availability "barriers" is necessary to work individually or in groups with high concentration modes.

Especially in this period of distance working, which has had the side effect of breaking down the already weak pre-existing barriers. How can this social habit take root?

Think about what drives you to meet your colleagues in the company. You will certainly think about problems to be solved in different functional areas and opportunities of a different nature.

Although it is possible to foresee well in advance that specific meetings will have to be held to deal with the various business needs, every year in all companies large amounts of time are lost in scheduling meetings at the last minute, chasing the availability of the various participants, interrupting ongoing activities, and in fact establishing the practice of continuously working in a fragmented mode – that is moving from one topic to another, highly variable and marked by a constant feeling of urgency.

Exploiting the principle that managerial work is divided between "one-off" elements—unpredictable and therefore difficult to organize a priori—and recurring elements that are repeated over time—both at individual and group level—the smartest thing you can do in your company is to organize a priori most of the meetings you can foresee. With a horizon of 6 or 12 months, cadence the various meetings and position them so that you no longer spend any time on their scheduling, except for exceptions, and above all try to homogenize the meetings by subject area.

For example, in several companies followed in recent years, the practice has arisen of dedicating one day a week, usually a Friday, for management meetings to review the closing week on the various fronts, for example production, sales, cost analysis, to set priorities for the following week. In other companies, it has become the practice to establish the management calendar for the whole of the following year in November-December of the current year. This calendar will include the monthly follow-up meetings of the strategic projects launched, the periodic meetings of the various defined inter-functional committees, the periodic "1 to 1" meetings between the various levels of management, and the periodic meetings of production, sales, marketing, and so on. We will deal with this topic in more detail in Chapter 9 on Lean Meetings. Making the annual calendar is a demanding activity, because in fact it cannot be done without the design of the company's operational governance model. But the effort is always rewarded with the clarity that follows and the simplification in everyone's planning, with company commitments known a year in advance. A further benefit is that having one big single representation of the various company meetings facilitates the identification of both redundancies and shortcomings. Moreover, the work done is inevitably improved from year to year and evolves hand in hand with the agile management skills of the company's managers.

Lean Lifestyle Story: Sammontana

Method, Routine, and Guide to Behavior in Innovation Projects

Sammontana, historic Italian food company specializing in the production of ice cream and frozen croissants, has launched an important initiative focused on exploration, generation, development, and market launch of new high-impact innovative solutions. Starting in 2018, an extended group of people was exposed to a new way of working and innovating with the "problem" of continuing to carry out all other activities of the day-to-day running of the business. How to keep the focus on the company's strategic projects, without getting "pulled" by day-to-day urgencies? The key to not losing any innovation project along the way was the introduction of a working method and organizational routines that gave strict cadences to weekly project appointments and monthly review appointments with the involvement of multidisciplinary teams from the very early stages of any project launched, called upon to play a role not as mere executors, but as co-leaders in the elaboration of direction input.

"We have always made innovation," says **Leonardo Bagnoli**, CEO of Sammontana, "but the partnership with Lenovys allowed us to acquire for the first time an agile working method to develop new products with a high impact on the market and for the company: this was the great added value."

From the point of view of organizational routines, we have worked to be able to focus the innovation team on innovation activities without being able to detach them from their own operations. In other words, the first challenge to overcome was to free up time to dedicate to innovation while continuing to guarantee supervision and results in day-to-day activities. The following elements were therefore put in place to streamline and simplify activities, ensuring rapid progress in an efficient manner:

- **Cadence:** the joint progress meetings where the whole innovation team, with Lenovys support, would meet from the beginning to the end of the project, were defined immediately. This made the intermediate deadlines clear and created a sense of urgency toward the expected results for those deadlines. Innovation activities also began to have a sense of urgency as day-by-day activities.
- **Availability Windows and Slotting:** time slots were defined early on throughout the two-month project duration available for innovation activities, synchronized with the availability of the different project

participants. This made it possible to structure the daily work in advance to obtain the necessary time slots to work on the innovation project.

■ **Continuous experimentation:** this principle, which is fundamental in a Lean Lifestyle Company, was applied both to the way of working, not to be subjected to a "closed box" but to be tested, experimented, and calibrated ad hoc, and to the same way of innovating products and proposing them on the market. Instead of waiting to complete the entire process of development, industrialization and launch, teams were guided through the rapid design and implementation of tests of all kinds to validate the different hypotheses before proceeding with development. This prevented solutions that would be blocked later, or from adjusting the shot well in advance of the traditional approach.

Sammontana's commitment to removing obstacles that might not favor good practices in the management of innovation projects has been maximum and has also concerned those consolidated behaviors that are difficult to eradicate. In this sense, a particularly significant example of action on functional innovation behaviors is that of the Impact Innovation Workshops. During these workshops, in fact, dysfunctional behaviors and the corresponding functional behaviors were stated at the beginning of the activities, to maximize the probability of generating high-impact innovation ideas.

The task of the workshop moderator was also to give feedback on the application of functional/dysfunctional behaviors, to eradicate the latter in favor of the setting of the former.

The classic "it can't be done" was replaced by the question "what can be done to make the solution feasible?" and still every time people rushed to investigate the feasibility of the idea, they were reminded to first investigate the value of a proposal to the customer and then its feasibility. It is indeed unproductive to spend hours of a meeting discussing the feasibility of a solution that ultimately has little value to the customer!

This way of working on behaviors has generated rather amusing dynamics within the working groups, where people noticed each other's functional or dysfunctional behaviors.

"The enthusiasm and positive energy generated by the projects conducted in the manner brought by Lenovys are very high," comments Bagnoli, "and the result of the market launches of the new products have also paid off in terms of business performance."

Knowledge Management

Why do we interrupt our colleagues or get interrupted by them during any activity? Let us think about it. Often the answer is in that information to be given or received or in that answer that one needs at that moment, and for which one is not willing to wait. Not least because one would not know what and when to wait. This is a problem felt in all companies, which have always grappled with different solutions to promote this communication in questionable ways: offices with disparate layouts, open spaces, information exchange areas, hundreds of e-mails, myriads of in-person and remote meetings. In the Covid-19 era, as we can no longer attend our crowded workplaces in person, we are populating the web and the various communication platforms with our virtual omnipresence. I do not want to be misunderstood. It is all well and good to foster communication between people in the company, whether in physical presence or at a distance. But the key point, from my point of view, is not to disproportionately foster communication between people but to organize all the information that people accumulate during their work experience and make it available to other colleagues in an easily usable and structured way, moving away from just "personal" logics and embracing the logic of building corporate systems. Information management thus evolves into the management, consolidation, and capitalization of corporate knowledge.

The accumulated know-how should every day go from the "heads of the people" to the corporate system daily, to convert "implicit" knowledge into "explicit" knowledge. Uncoded knowledge into codified knowledge.

Why should this transition promote fluidity and simplification in the company? And why should it reduce interruptions? Imagine that you want to know whether a purchase request of yours has been processed or not. You have two choices. Either contact your colleague in person who oversees managing the purchasing process or press a button, at most two, and see where your request stands. Imagine you want to know whether an action agreed in last week's meeting has been done or not. Always two options: contact in person or two clicks if there is an up-to-date corporate digital system. Which would you prefer? Which would waste less of your time?

I did an experiment. I had to book a doctor's appointment. I had two alternatives. One was the traditional way, by phone and in person, the other was the digital way, using a "cold" online booking platform. In the first case, even though it was a private and not a public medical center, I had to wait many minutes on the phone before being greeted by a secretary to whom

I asked several questions about the doctors available to make the visit I had in mind. She was very familiar with the doctors' agendas but did not know the differences in detail between one and the other, claiming that they were all excellent professionals. I booked a visit after about 20 minutes for an amount of 153 € to be paid at the time of the service. I completed the experiment with the second alternative. I used a private booking platform structured as an aggregator and web comparator of health services and benefits, totally free of charge, which promised to "check in real time the agendas and availability of private and non-profit facilities and allows you to search, book, pay, a social-health service with just three clicks, avoiding queues and unnecessary waiting."

In just 2 minutes I booked the same visit as before, after also reviewing the detailed CVs of the doctors I chose, paid by credit card, avoiding further waiting at the time of the visit. I forgot: €132 versus €153 for the same booking as before. Inexplicable difference, also considering the additional commercial commission costs that the facility will have to pay to the platform. I recalled the medical center and cancelled the phone reservation made a few minutes before. The secretary, without much ado, canceled the reservation and did not ask me any questions about the reason for my request for cancellation.

After I had overcome my initial hesitation, digital service has overturned the challenge. I wasted much less time, saved money, got more information and above all, easily found what I was looking for.

Imagine the same kind of benefits in your company. Requests for information and updates on ongoing projects, initiatives, action plans and problem solving. Exploring past problems and solutions, lessons learnt, useful for dealing with a similar problem right now. Requests for "first aid" on various topics. Transmission of information to anyone who may need it or who needs to be informed.

Today, we have an embarrassing amount of technology options to collect and make available information and know-how: from advanced PLM systems to manage all information and workflows during product development, the dozens of workflow and collaboration management solutions, via the various document management solutions.

How do we choose the solution that suits us? **The advice is to start not from the existing solutions on the market, but from the analysis of internal processes and the real criticalities experienced in your organization.** The objective must be to simplify/fluidify on the one hand and to increase the value of the company on the other. Because when the know-how is decoded, managed, and made easily usable, becomes tangible assets of the company, contributing to its growth in value and performance on the market.

A completely digitized corporate Knowledge Management system will represent a real "working handbook" for all employees, with different levels of access. An operational center of all information necessary for the functioning of the entire company, structured to facilitate quick and easy access to those who need it and at the same time allow the creation and integration of new content, by all company contributors.

Checklists and FAQs

Every day in any company in the world, countless tasks are performed. Faced with any task, everyone organizes and performs the required activity in his or her own way, inside and outside the company. Some things that are natural and obvious to me will not be so to my colleague. It is normal. In a Lean organization, using checklists and procedures is the most natural way to define a shared standard of work and then trigger the continuous improvement process.

This principle does not only apply to work operations shared by several people. It also applies to complex operations carried out several times by a single individual and which we want to be carried out with zero errors.

Imagine an airline pilot and his crew. I don't think you can doubt that their performance should always be error-free. It is no coincidence that on any scheduled flight you will always see thorough checks carried out before each take-off by means of checklists.

Instead of proceeding as it happens and maybe forgetting one thing or another, checklists ensure that all systems are functioning according to standards and that lessons learnt from the past are translated into operational guidance for the future.

Checklists reduce the possibility of pilot error to almost zero. At the company, we have endless possibilities for introducing checklists to ensure consistent high performance, zero errors and zero subsequent interruptions. An example is the "engineering checklists" used in all phases of new product and service development.[2] In the chapters on Lean Mailing and Lean Meeting, we will look at other examples of the application of company checklists.

A close relative of checklists is the tool of FAQ—frequently asked questions—necessary to reduce the need to provide information that may be of interest to several people and that can be predefined in advance. This is exactly what many insurance and banking companies have done, transforming their call centers into a sophisticated digital FAQ system to deal with all predictable customer queries before interrupting and wasting an operator's time with information that can be conveyed in a more streamlined manner. You can design and embed FAQs and tutorials wherever in your company information needs to be provided in advance, to reduce interruptions and non-value-added work for everyone.

Notes

1. Gerald Weinberg, *Quality Software Management: Systems Thinking*. Dorset House, 1991.
2. Luciano Attolico, *Lean Innovation. Strategies for Enhancing People, Processes, and Products*. Hoepli, 2012.

Chapter 8

Halve E-Mails in the Company and Double Their Effectiveness: Lean Mailing

You Already Know

After many minor jobs, I started working in a big company as a new engineer in 1995, in Magneti Marelli. One of the nicest memories of that time concerns a scene that was repeated every day: the distribution of the internal mail by Giuseppe, who was also responsible for photocopying mountains of paper for internal use. Giuseppe delivered internal mail twice a day, even though we already had electronic mail. Important information, despite the new technology, was transferred by Giuseppe in the morning or afternoon. Although there was a "swarming" of communication between all of us, I do not remember the constant interruptions typical of the "way of working" of our days.

The question I often ask myself is in fact: "In the last 25 years, have we gained efficiency in communication and exchange of business information or not?"

My first answer would be: yes! However, by stopping to reflect and observe better, what has happened touches deeper spheres of our way of working and the picture is much darker. While it is true that we have witnessed an irresistible rise in the efficiency of technological tools that have become so powerful in transmitting almost instantaneously massive amounts of information around the globe, it is also true that there has not been an equivalent rise in the effectiveness of communication. And,

DOI: 10.4324/9781003474852-8

unfortunately, the total bill we are paying socially—in terms of real productivity and lifestyle—is a steep one. In fact, in addition to the dubious quality of the information circulating, we have internalized both at individual and organizational level a series of dysfunctional behaviors that have strong negative impacts in terms of the overall time spent. A bit like what happens with various drugs, we treat something immediately, but in many cases, we suffer side effects that we realize only later. **E-mail was originally a formidable tool**. Over time, its use has become distorted on a professional level. We are obsessed with compulsively checking e-mails in real time and find ourselves unable to stay focused in a meeting without checking—several times—all the e-mails and messages we have received. We work in organizations that do very little to bring this tool back to its original purpose: to transmit, wherever there is a connection available, important information to targeted and selected recipients, in any corner of the world, at almost infinite speed, without, however, expecting instant replies. A great and efficient evolution of traditional mail and fax, a perfect system of asynchronous communication, in which question and answer are deferred in time, which unfortunately in common, distorted usage, people try to make synchronous and intrusive, cancelling the distance between question and answer and involving more and more people in communication than the minimum necessary. We witness the continuous increase in the volume of circulating e-mails. We cannot "unplug" wherever we are, and we accept— *obtorto collo*—an additional silent source of stress in our lives, which we cannot easily get rid of. In addition to the insult, the world of e-mail represents a cost item in constant rise in most companies.

In Depth

On average, a modern worker spends **5 to 20 hours a week** reading and managing his or her e-mails. Out of around **two hundred e-mails received every day**, only between 10 and 20% turn out to be useful. Currently, the numbers associated with the circulation of e-mails have taken on embarrassing proportions and with very heavy consequences for people and companies: worldwide, at the end of 2019, there were 3.9 billion e-mail users and **around 293 billion messages sent and received every day**. And by 2022, it is estimated that users will grow to 4.4 billion and the number of e-mails to 347 billion.[1]

These are numbers to keep an eye on, especially if you consider that, according to some studies, including those of Ademe, the French Environment and Energy Management Agency, 8 e-mails pollute as much as 1 km by car and that a single 1-megabyte e-mail can emit up to 19 grams of carbon dioxide.[2]

Increased e-mail volume and increased smartphone usage time go hand in hand. In fact, according to research carried out by Adobe, 85% of users check their mail via smartphone and if we consider users between 25 and 34 years old, the percentage rises to 90%. On the other hand, several studies have shown that we spend an average of eight to ten hours a week dealing with incoming e-mails, beyond working hours.[3]

All this is leading to negative effects on productivity and performance that can no longer be overlooked by entrepreneurs and managers. Under the heading of *Lean Mailing*, I collected, together with my team, the best individual and corporate practices to re-learn how to manage e-mail with the maximum results and minimum waste of time and resources. Through Lean Mailing, individuals and companies discover that it is possible to gain time and energy wasted unnecessarily due to the failure to apply rules that are as simple as they are effective in e-mail management.

Here, in brief, is the typical structure of a Lean Mailing path:

1. **Value**: understanding where the value is and where the waste is in e-mail management
2. **Elimination and reduction**: eliminate unnecessary e-mails, reduce the quantity and size of everything else
3. **Effectiveness**: when needed, write e-mails that work better
4. **Individual habits**: review when and how to manage e-mails in the day
5. **Business processes and systems**: reviewing processes, systems, and rules to improve e-mail management
6. **Continuous Improvement in the corporate e-mail system**: establish specific business metrics and their monitoring.

Value

Have you ever received an e-mail with 20 lines of text without a single wrapper? Or with an attachment so heavy that it is impossible to download with the hotel's Wi-Fi connection? Or one that had undergone "some" forwarding in the last 2 months, becoming a "chain letter"? Or an e-mail with 13 other people copying it? Or with a task assigned to you, to be closed by yesterday? Or containing a direct link to a page or a 50-page document, where you don't know exactly what you must read? Or an e-mail that doesn't explain why they wrote to you?

Do you think in similar cases you create value with the use of e-mail, or is it just a waste of your time and that of others?

Each e-mail must be written and sent only if there are no better alternatives to it: telephone call, message, chat, online update of the status of a project or a document shared in the network or physical update of a visual management system, etc.

Once I have established that I *must* write an e-mail, this should be conceived to add value to my colleague-internal or external customer. Yes, in the Lean world, the colleague is also a customer, and I am his supplier. I must make sure I give him value through every interaction. No redundancies and frills. Extreme care in summarizing and the key messages I want to convey. Be careful not to put unnecessary people among the direct recipients, extreme care in the use of knowledge copy recipients, avoid indiscriminate use of the embarrassing *bcc*. If you want something from someone, be brief, specific, and clear, leaving no room for interpretation and unnecessary rounds of subsequent e-mails.

Never use the e-mail tool to negotiate, to find a date that suits everyone, to discuss technical proposals, to do root cause analysis, etc. It is good to use e-mail to get timely answers, to formalize agreements, to assign tasks. One-time bickering. Not a continuous repartee.

Elimination and Reduction

Today, the various sources of incoming messages have increased by leaps and bounds (e-mail, social network feeds, chat, WhatsApp, etc.) and the quantities of messages for each type of source. Far too much. Using e-mail, social networks, and everything else, is unfortunately something that nobody has taught us how to use well. The first thing we must learn to do is to eliminate without delay not only the mountain of superfluous communications that storm us daily, but also all the useless actions we take to manage them.

Let's see how:

- We cancel subscriptions to all mailing lists where we are registered and have not used them in the last three months.
- We delete incoming e-mails where we are acquainted with them, as well as those we have already read, and don't even waste time archive them. Delete them. Unless they are documentation, certifications, or official communications that you are "forced" to keep.

- Let's get used to the question: "Will I certainly use this information for something immediate and important?" If the answer is NO, let's not keep it!
- Let's also stop writing e-mails that we could avoid with a phone call, a message, an update on a document shared on the Network available to your colleagues. Even better still with a meeting in person.
- We expressly ask not to be made aware of anything that might concern us or be of interest to us. We will ask for the information we want when we need it. If someone wants something from us, let's make sure that we are specifically asked. If we are the ones who want something from someone let us, ask them in an equally direct and specific way.
- Let us stop using the "cc" in the rain hoping that someone involved does something for us or even thinking about empowering the group. Distributed responsibility does not exist precise tasks go to precise people. Let's also stop thinking about putting "pressure" on the recipient by showing them who else we are forwarding that e-mail to for knowledge. At most, let's write it in the text of the e-mail, but let's save the easy forwarding to people whose work is unaffected by receiving that e-mail. Let us only put in cc's those who need that information to continue their work or who otherwise risk making mistakes or wrong choices.

Effectiveness

We often write our e-mails out of the box thinking we can buy time. We usually lose much more of it afterwards than we initially gained, because we rush through the misinterpretations, the need for clarification, and other e-mails answering questions that could have been avoided with a more effective one. All rework work, redundant and unnecessary that falls on us and those who receive our e-mails, when not written with the maximum possible effectiveness in mind. The more effectively an e-mail is written, the less time it will take to read it and to get straight to the point without wasting time. This applies both when an action is required and when the purpose is purely informative. Here are some criteria for increasing the effectiveness of our e-mails.

The Object

When we write it, we must always keep in mind that almost everyone now receives tens or hundreds of e-mails a day. So, the object must make our message recognizable on the fly for what it really contains, it must already be a message in miniature, like the title of a news article. We give our interlocutor the freedom to go deeper into the matter, reading the details by opening the e-mail, or not to open it. **No, therefore, to generic words and vague expressions**, especially if the recipient of your message reads the e-mail via a smartphone. When the reader receives your e-mail, he or she must immediately know what to expect. The short, clear subject line that gets straight to the point, which anticipates the action or key information, is opened, and creates the habit of conscious openness.

The Text of the E-Mail

One mistake not to make is the easy digression. The content of the e-mail must be consistent with the subject and deal only with what you anticipated in the fateful first line. When we send a very long e-mail asking our interlocutor something, we simply show that we have no respect for his time. Try to write the shortest possible text and get the eventual answer as simply as possible. Try to get to the point without too much preamble, always remembering to write why you are writing, what you want and from whom, by when you want the answer. Better to have more content in the same missive or better write two or more separate e-mails? To optimize the processing of requests, it is ideal to send separate e-mails for separate topics! Especially if the themes require different processing times, if the urgency or the tone of the tasks is different. This allows the receiver to use the inbox as a task checklist, meaning that each closed task will correspond to an archived e-mail. If you just can't avoid sending that long e-mail you have in mind, let us at least remember to create a foreword that makes it easier to read: a quick summary, a small index, or a short list of topics. Always remember not to overload your interlocutor and bear in mind when starting to write the text of the e-mail: the text should consist of sentences not exceeding 3 lines to facilitate reading and comprehension.

    Also, pay attention to the questions we ask in the text.

    Ask them well, specifically, not generically open, and above all, ask only a few (tendentially one or two at most). Always ask specific questions that require simple and equally specific answers.

The Last Challenge: The Standardization

Horse wins, doesn't change! Why not create standard formats for our recurring e-mails? Creating templates will help us reduce several time wasters that are imperceptible in the moment when I live it, but relevant in the long run when repeated several times. Try multiplying just 5 minutes of time by a couple of times a week, for a year. Have you counted? That's 500 minutes, or just over 8 hours in a year. A whole day's work for that trivial e-mail, which, multiplied by 10 years, makes 10 days of your life go up in smoke. But I am convinced that it is not just 5 minutes and that the days burned in 10 years are many more. But unfortunately, we do not realize this. So what? We can start by asking ourselves what type of e-mails we send most frequently. For example, for those who travel a lot the answer may be "hotel bookings" and, if I don't simplify with a phone call, I can create a template that serves this purpose. There can be standard e-mails of meeting minutes, standard e-mails of departmental communications, where the structure remains the same and only the data changes. Keep thinking about it. I am sure that with a little creativity you can come up with many other examples of standard formats, for you and your colleagues.

In conclusion, about effectiveness, it is good to always remember to safeguard the time of others. However, it is even better to remember very well that every time we write more than strictly necessary, or with such sub-optimal effectiveness as to require clarification, rework, and further elaboration, it is mainly our time we are talking about. Of our most important resource. And we are the guardians of its precious use. No one else can do it for us.

Individual Habits

Our mailbox is the digital equivalent of a real mailbox. What do you do when you leave home in the morning and find correspondence in your mailbox?

Most likely, you take the envelopes, open them, read them, throw away most of everything you have received, put only what you want to read at another time or everything you want to reply to in your bag. The letterbox remains empty and ready to receive something new in the hours and days to come. What you do not is read the incoming communications and then put them back in the letterbox! Because then you would find within a few weeks your letterbox full of useless things mixed with a few important communications now lost in that mess. Think about your inbox and even more precisely your inbox: is it empty right now? Will it be tonight or tomorrow? When? What habits have you established about it?

I think we all felt that bad feeling that there is some communication or some request waiting for an answer, hidden somewhere. A feeling that comes from the fact that we cannot easily get control of a tool that often puts us in a state of blatant out-of-control, not being able to govern, in almost all cases, when and how we receive the next e-mail. Since we cannot curb or change the flood of information, requests and communications that pours into our inbox every day, the only thing we can control in e-mail management is what we decide to do in relation to them, what reactions we want to have, when to read, when and how to write our e-mails. That is, what habits we choose to adopt in this regard.

We should not take it for granted that the current habits acquired in terms of e-mail management are those that ensure us the maximum result with minimum time spent. If you want to bring the management of e-mail inboxes and other messaging sources back under your control, and especially if you want to limit the damage of multitasking and reduce context switching times, rework, and wasted time and energy related to the mismanagement of e-mail and messaging in general, here are some concrete examples of behavioral habits that have proven to have great impact:

- **Switch off the sound signals** of receiving e-mails and various messaging, if possible, always, but absolutely when you are in sacred time mode—high concentration—alone or in a working group.
- **Never read e-mails as the first action in the morning** (unless your job is to answer e-mails, e.g. customer service, customer care, call center, etc.). You end up chasing other people's priorities, but not your own, and induce a reactive and unfocused state of mind.
- **Set specific**, highly concentration **time slots** to read and handle e-mails. A frequency of 2–3 times a day is sufficient for almost the entire professional population. With just one slot a day, one survives greatly, provided one overcomes initial withdrawal.
- Process e-mails and various messages, during the dedicated e-mail handling slot, with a **standard routine to process all** received **e-mails:**
  - delete directly, without opening, mail messages falling into the categories already seen in the previous paragraph;
  - open e-mails, read, reply immediately or delegate/forward in all cases where this action takes less than 2 minutes;
  - put a flag or a reminder when you foresee more time needed and must therefore postpone the relevant reading or response to another scheduled slot;
  - delete or archive everything after these individual steps;

– don't waste time on useless archives and folders because you only need one, as search technology will now find everything you want when you need it.

■ **Apply the rules of effectiveness** seen before to every e-mail you write.
■ **Avoid reading e-mails in the last hour of the evening**, before going to bed.
■ **Plan 1 day a week completely free of e-mails and messages**.
■ **Communicate to colleagues how to handle e-mails**. This will avoid misunderstandings, promote alternative ways of communication and above all set a concrete example for a new way of working. A new way in which e-mails and messages bring greater effectiveness and efficiency, only if used correctly.

If you have any doubts about these new habits, let me ask you a couple of questions:

When and how did you design your current habits?
And *for what purpose?*
Are you struggling to find the answers? It doesn't matter. For me, the objective is clear. I am talking about the preservation of our time and energy.
How? The many successful experiences of hundreds of managers who have tried before us show that it is not only possible, but necessary to do so.
When? A very simple answer: now! And if you can think of any other key habits not mentioned here, write to me, and let me know. Write to: luciano.attolico@lenovys.com.
Subject: "Lean Mailing tip from a Lean Lifestyle book reader."

When I started working by slotting in e-mail management this immediately freed up my time because I reduced my e-mail time by 25 per cent. At first, I was a bit anxious to check if someone had written or to find myself swamped with e-mails to read at the next slot. But then when you see that it doesn't happen you start to use the time you saved profitably. I'm happy to have taken the urgency out of answering everyone immediately and I also see that it has caught on in the company.

Daniele Regazzi, R&D Manager Lucchini RS

Business Processes and Systems

I often meet entrepreneurs and managers who complain about the way their employees work, and how we are now flooded with e-mail and various messaging. On the one hand there are nostalgic feelings toward times long gone—only 10–15 years ago—when no one was as helpless as today under the current deluge of messages, but on the other hand I see no corporate planning to provide concrete alternatives. Companies have a great responsibility in setting up processes and rules to improve the e-mail management of individuals.

Let's see some examples of company-wide structured actions for this purpose.

- **Design an e-mail- and message-free slot for everyone.** For example, one afternoon a week, within a defined time frame, could be the right signal to draw everyone's attention to the issue and give the company a few hours of highly concentrated work that we are no longer used to.
- **Limit the amount and weight of e-mails that can be sent per day.** Strangely enough, this action leads to unexpected results in terms of reducing the number of e-mails circulating and increasing the effectiveness of corporate communication, because people are talking to each other again.
- **Limit the number of people in the company to copy each e-mail sent.** This action drastically reduces unnecessary e-mails circulating in the company.
- **Prohibit consultation of e-mails and messages during meetings.** In many companies this concept has been acquired as "Far West meetings" or "smartphone guns" outside the saloon.
- **Define and disseminate a set of common business rules**, for example:
  - the e-mail starts with INFO when it is information that does not require a reply;
  - facilitate the identification of specific actions required of recipients by directly indicating them with "@ PERSON—action—by when";
  - the recipient in CC does not have to reply to the e-mail;
  - write "END" at the bottom of the subject line in all e-mails that have only the subject line, so that it does not even get opened;
  - never ask for e-mail confirmation;
  - never call and say: "I sent you an e-mail";

- update the title of the e-mail if the objective or focus of the topic changes;
- appointment management only via online calendar services, such as Outlook or Google Calendar, and never via e-mail;
- minimize attachments and require document sharing via links to corporate network folders.

Lean Lifestyle Story: Campari

Processes Matter

Talking about the number of e-mails or the number of phone calls we receive, and which often interrupt us, sometimes we forget that it does not depend only on how we are organized or how "others" decide to communicate but are sometimes the direct consequence of the business processes the company has chosen to set up.

We can see this correlation in a project addressed by the Lenovys team in Campari, the world's leading producer of soft drinks and alcoholic beverages. A team tackled the import/export process of "spirits" with the aim of streamlining it and increasing the level of service.

The team quickly identified some important wastes:

- many informal communications via e-mail and telephone
- Useless documentation
- messy archive with difficulty finding documents
- Repeated code changes in sales orders

Moreover, the process was certainly made complex by the lack of a single information system and because it required communication between many different actors, such as origin plants, destination plants, customer service and planning hubs.

Based on these reflections, the team defined the following as the main countermeasures:

- Digitization and organization of the archive
- Reorganization (using the 5S Lean method) of the archive that was to remain on paper
- Standardization or optimization of the process flow

The digitization of the archive has positively influenced document search times, but what made the difference were the redesign and standardization of the process that had a strong effect on the information flow and thus on the number of e-mails and phone calls required to handle an order, ultimately impacting the overall workload.

At the end of the redesign, the team achieved a 67% reduction in the overall workload of the business and reduced the flow of e-mails related to process management by 87% and practically reset the phone calls, as shown in Figure 8.1.

Incidentally, in redesigning the process, the team also noticed specific waste generating extra costs in the container transport process that had never been directly correlated with the import/export management process: by optimizing and standardizing this management, they were therefore able to generate an additional €80,000/year in savings.

Luca Saporetti, supply chain Campari director in the period of realization of this plan and the current global vice president supply chain of the multinational, tells:

> This example, the first in a series, has once again highlighted the potential to apply Lean methodologies to processes that are less common and therefore so virgin in terms of improvement opportunities.

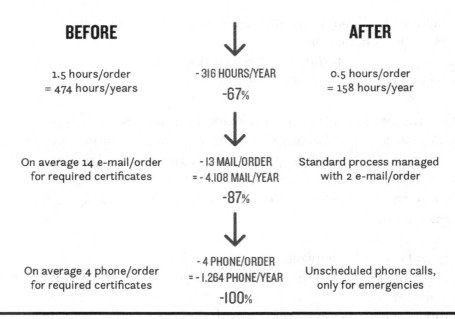

Figure 8.1   Results of the project to simplify the import/export process.

Lean Lifestyle Story: ELT Group

Lean Mailing and High Concentration

For 70 years, ELT Group has been a high-tech company that designs, develops, and manufactures systems of strategic surveillance, defense, and electronic countermeasures for naval, air and land use. It supplies the armed forces and governments of 30 countries with over 3000 high-tech systems. It is on board all major modern military platforms such as the Tornado fighter, the Eurofighter Typhoon, the NFH-90 helicopter, the Italian PPA platform, and the Italian and French ships Horizon and FREMM.

The company is part of the ELT Group, which also includes CY4GATE, specializing in Cyber EW, Cybersecurity, and Intelligence, and ELT Gmbh, a German subsidiary specializing in the design of Homeland Security systems.

A company with great momentum and a very high interest in the full development of the potential of its more than 900 employees, including through the encouragement of agile working methods.

When the first survey on the dissemination of Lean Lifestyle principles in the company was carried out in 2017, among the most critical areas highlighted were e-mail and high concentration work. A pilot improvement project in the Design Solutions department for the application of Lean Mailing and the reduction of interruptions has therefore started, in order to increase the percentage of high concentration work.

The initial mapping identified and quantified waste, highlighting the following main problems:

- E-mail with subject 1 or unclear text;
- E-mails sent or forwarded to unnecessary recipients (e.g. excessive cc);
- Work interruptions related to reading e-mail or notifications;
- Distractions of those working in the open space related to noise or conversations;
- Interruptions due to telephone calls;
- Interruptions by colleagues or bosses.

The main improvement actions were:

- Dissemination to the entire department (60 people involved) of the habit of reading e-mails in predetermined slots with initial training, design of the new habit and daily monitoring of success in observing the habit.
- Elimination of e-mail notifications from PCs and mobile phones.

■ Reduction of environmental obstacles that generated interruptions. For example, the "table meetings," short operational meetings of 2–3 people in front of a PC, which were very effective but generated noise and distraction in the open space, were diverted to adjacent areas in non-bookable mini-meeting rooms specially arranged with high tables.

■ Three information campaigns on the correct use of e-mail: on recipients, on how to write the subject, and on how to write the text. In particular, the one that explained which rules to use to define e-mail recipients, which led to a 32% reduction in forwarded messages and "broadcast" e-mails, i.e. those sent to many recipients.

■ Creation of a feedback system on the quality of circulating e-mails with respect to the themes of current information campaigns, which resulted in a total of more than 3,000 feedback sent in 5 months of the project.

■ Monitoring, feedback, and discussion of results in monthly meetings with the entire team (Figure 8.4).

As an effect of the project, all types of interruptions were significantly reduced as shown in Figure 8.2.

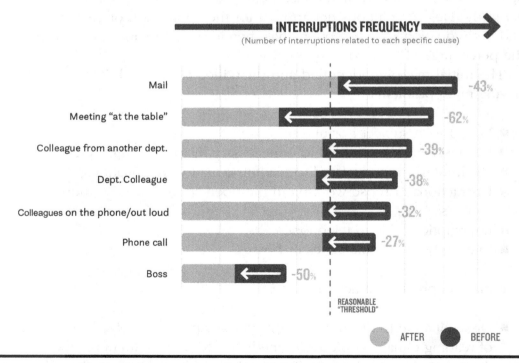

Figure 8.2 Results of the interruption reduction project in ELT Group.

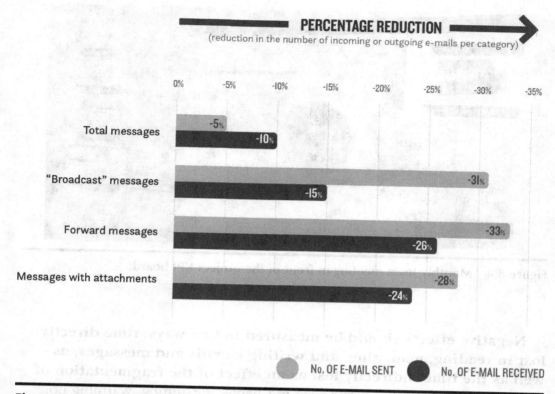

Figure 8.3   E-mail reduction project results in ELT Group.

The number of circulating e-mails also experienced a significant reduction, mainly focused on forwarded and "broadcast" e-mails as shown in Figure 8.3.

Continuous Improvement in the Corporate E-Mail System

You don't improve what you don't measure. In the world of work people have no idea how much time can be destroyed each year due to the mismanagement of e-mails and various messages, and above all, almost everyone confuses reading and managing the daily flood of e-mails and messages with the work itself. In Lean logic, we should learn to consider this area of work as a form of "necessary waste." Not disposable, but to be minimized.

Figure 8.4 Monthly team meeting in front of the project KPI board.

Negative effects should be measured in two ways: time directly lost in reading, managing, and writing e-mails and messages, as well as the time indirectly lost as an effect of the fragmentation of the actual work, in the forms seen in Chapter 4: (shifting, warming up, and bad quality time).

If we want to trigger virtuous cycles of continuous improvement in this area, company-specific training courses on the subject must be designed, followed by structured improvement projects at individual and organizational level. Each path starts from the identification and sharing of the reason for acting, which usually includes both elements of corporate communication ineffectiveness and inefficiency, with the quantification of the direct and indirect effects of lost time, as well as undeclared costs incurred because of dysfunctional behavior. Measurement is fundamental. This is why a quantitative and objective assessment is carried out to have a numerical basis with which to compare during the project. Today, many applications are available that can be used for this purpose: "Bells & Whistles," "Outlook Statistics," "ReliefJet," "E-mail Meter," and "Outlook Statview" are just a few examples of Apps that you can use, together with your IT department or whoever follows corporate information systems for you, to get an initial frame of reference and build a monitoring system during the project. Immediately after defining the targets of the results you want to achieve, we move on to the analysis (1) of the business processes that can be modified to achieve the desired results and (2) of the key behaviors that need to be redefined to ensure the success of the program. Behavioral analysis needs monitoring and continuous follow-up feedback that underpins progress. For these monitoring and feedback activities, appropriate "change agents" are usually selected and trained to become real "enzymes" to digest the new processes and behaviors within the organization.

Lean Lifestyle Story: Lucchini RS

Personal Excellence and Lean Mailing

When one wants to spread the principles of the Lean Lifestyle, one of the first fundamental steps is to combine the company-wide actions with extremely practical training programs in order to act simultaneously on the individual habits and beliefs of individuals and, in particular, managers.

The structure of these training programs is typically made up of five key moments in order to effectively generate changes in people.

1. **High-impact initial training**, aiming not only to spread knowledge of the principles and operational tools of the Lean Lifestyle, but above all to create a strong awareness of the problems and opportunities.
2. **Launch of an individual project by** each participant focused on the application of the ATRED method to redesign their way of "functioning" and interpreting their organizational role work, with an indicative duration of 3–5 months.
3. **One-to-one mentoring and coaching**, which assists each person in the development of his or her project through a mixture of three elements: support with the **coaching approach**, push for change with the **nudging approach**[4] and guidance with an experienced mentor who has "walked that road before."
4. **Explicit moments of** individual or collective **reflection and learning** during or at the end of the course (*hansei*).
5. **Monitoring and integration at company level, so that on the** one hand the results of the application are kept under control, and on the other that they can be integrated into company processes and habits (e.g. introducing ATRED[5] as a standard method that everyone uses).

A program with this structure, for example, was launched in 2018 at Lucchini RS focusing, within the framework of the Lean Lifestyle principles, it was focused on issues related to Personal Excellence in the company.

The program has so far involved 63 managers, including the CEO and his front lines, and by the end of 2021 will involve another 25.

As an example, it is interesting to observe the improvements that the lean mailing approach has already achieved by working on an individual project level.

Surely the time spent on e-mails is always one of the "sore points" of all the participants and therefore many of the actions focus on this, with two objectives: on the one hand, to reduce the total time spent on this activity and, on the other hand, to eliminate stress, fatigue, and pervasiveness in work and private life that e-mail multitasking often causes. Figure 8.5 shows a statistic on the group of participants of the recently concluded edition— February 2021: twelve participants, both managerial and operational, and from various backgrounds (design, research and development, planning, commercial, maintenance, and production). First, we can appreciate the

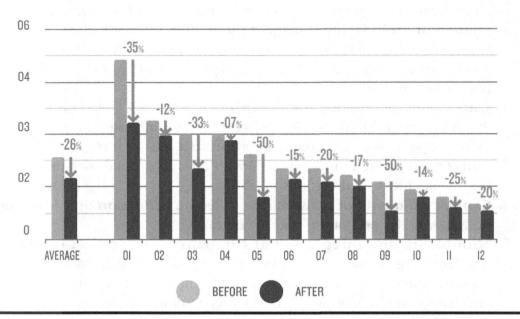

HOURS PER DAY DEDICATED TO E-MAIL MANAGEMENT AND IMPROVEMENT AFTER AN INDIVIDUAL LEAN LIFESTYLE PROJECT

Figure 8.5   Results of the Lean Mailing project in Lucchini RS.

extreme variability of the daily time spent on e-mails, which apart from a few "outliers" fluctuates a lot (more than +/-50%) around the average value of almost 2.3 hours/day. While there are differences in the type of function belonging (e.g. sales vs. production) or role (e.g. manager vs. operative), experience shows that very often the variability is strongly influenced by the individual, their managerial/operational choices, and the approaches decided to follow.

In the graph, you can appreciate the results of the efforts made on this specific item of personal ATRED.

The average improvement was 26% with peaks of up to 30–50% and only one case being less than 12%.

As can be seen, the improvement is also quite independent of the number of daily hours of departure, even if for experience below 30–45 minutes/day the percentage improvements tend to decrease.

How did they do it?

Let's see, with a summary of the most recurring countermeasures taken, the main actions:

- **Deleting notifications** from PCs and smartphones;
- **Mail slotting**, each with its own reading frequencies and durations depending on role and function;
- **Batching "bureaucratic" activities** notified by e-mail (e.g. approvals, RDAs, and signatures);
- Questioning of **archiving methods** (you have no idea how much time some people invest in very detailed archiving that they sometimes use only a few times a year!);
- **"Structured" fight against e-mails in cc;**
- **Induced effects of delegation** on the number of e-mails received or reading times;
- **Communicating and training** others on the principles of Lean Mailing.

Eliminating the effect of multitasking is very often a harbinger of another important side benefit of this type of improvement: strong reduction of "instant response" stress to e-mails, even at home.

The opportunities are high. For this group of twelve people, the total time spent each day on e-mails was about 28 hours and the reduction achieved is equivalent to an overall saving of 7.1 hours/day: practically one working day for an individual.

The extension of the program and the systematization of the solutions, with structured interventions for training and to set new habits of Lean Mailing at the company level, are the necessary steps to amplify these improvements to the entire organization.

Notes

1. www.statista.com/statistics/255080/number-of-e-mail-users-worldwide/
2. www.green.it/le-e-mail-inquinano/
3. For further reading: The Radicati Group, Inc. E-mail Statistics Report 2015–2019 and 2017–2021 (www.radicati.com) and https://info.templafy.com/blog/how-many-e-mails-are-sent-every-day-top-e-mail-statistics-your-business-needs-to-know
4. If you want to learn more about nudging you can read the wonderful book Nudge: Richard Thaler and Cass Sunstein, *Improving Decisions about Health, Wealth, and Happiness*. Penguin Books, 2009.
5. The ATRED method will be analyzed in Chapter 10.

Chapter 9

Halve Company Meetings and Double Their Effectiveness: Lean Meetings

Which is the activity that consumes the most time and resources in a company? The battle to win this challenge has only two contenders: e-mails and meetings. According to research published by the *Wall Street Journal*[1] in 2019, there is no doubt: meetings win hands down. According to the data reported, the time spent in meetings is on average around 20 hours per week, between 30% and 50% of one's working time. And unfortunately, it is not always about high-quality hours.

Think about it for a moment.

How many times did you do something else during a meeting?

How many times have you answered your phone simply to say you were busy or occupied in a meeting?

How many times have you simply thought about something else instead of what the person on duty was saying?

And how many times have you seen your colleagues behave the same way?

Besides the negative impacts on your individual productivity, non-essential or unproductive meetings cost companies a stratospheric amount of money: $37 billion a year in the US alone.[2]

Do you have trouble imagining these important figures? Let's take a small-scale example and imagine that you have a company with 100 employees, including 10 managers and 90 collaborators.

DOI: 10.4324/9781003474852-9

Managers, at best, spend 16 hours per week in meetings, while employees spend 5 hours on this activity. Assuming 45 working weeks per year, an average hourly cost for managers of 40 €/h and 25 €/h for co-workers, we arrive at an expense for the company of approximately 800,000 €/year for the "meeting" activity. Imagine now optimizing and reducing the overall time by 25%: you will have saved 200,000 €/year for your company. Do you think it's impossible? By eliminating meeting start-up delays you will have already reduced that cost by about 10%. If you end up on time, the number of participants is reduced to the indispensable ones and eliminate a few excess meetings, you won't struggle to exceed those percentages.

The point is that not only will you have saved a lot of money, but you will have gained valuable hours for real work that no longer must wait until late at night or Saturday morning or who knows when, to be completed. **Rewriting the way of working necessarily requires a radical change in the way meetings are managed and conducted every day in the company**. Meetings themselves are not the problem. They are essential for teams and organizations, supporting inclusion and involvement, communication, and coordination. We cannot abolish meetings, online or in-presence, but we can make them shorter, more effective, and more efficient with the help of recent research on the subject and applying Lean Lifestyle principles.

Why Is It So Easy to Waste Time, Money, and Energy on Meetings?

It is paradoxical to see how in many companies there are *vade mecums* and rules written and displayed for the proper management of meetings. The point is that very often knowledge and common sense do not correspond to good practice.

We all know that we must eat a portion of fruit and vegetables three to five times a day, that we must exercise at least three to four times a week, that we must drastically reduce our consumption of saturated fats, sugar, and refined carbohydrates if we want to improve our physical and mental health. But who among us follows these good habits of writing a letter every day and every week?

As I said before, **there are two key factors that justify most situations in personal and professional life: decisions and habits**. It is these factors that enable us to sustain any change.

Unfortunately, most companies and most managers consider poor meetings inevitable, accept delays, meetings inevitable without decisions and action plans, constant interruptions and the use of PCs and telephones during work.

As if all this was imposed and decided from the outside! A bit like suffering a bad rainy day and not being able to do anything. But unlike the weather, meetings can be improved, if you make the right decisions and adopt the right habits.

If we accept justifications such as—what's 5–10 minutes late—excuse me I have to open the computer because it's urgent—I have to answer the phone because it's the boss—and I could go on and on, the difficulty is not in sharing a few rules for the proper running of a meeting, but in deciding what to accept and what not to accept in the company and changing the non-functional habits that have settled in the working groups.

Sometimes it is necessary to recognize that the problem starts with us. We are bad judges of our leadership skills in meetings. In a Verizon survey of more than 1,300 business executives,[3] 90% of meeting participants admit they thought about something else during the meeting, or not remembering large or small parts of the meeting; 79% of them reported that the meetings they led were extremely or very productive, but only 56% said the same about meetings initiated by others. Those running the meeting will have a perception of productivity that is not always aligned with those who are attending. Don't be afraid to ask your co-workers and colleagues what their perception of meeting quality is. You will discover interesting things.

There is not only a lack of perception of the problem. Often, we are also faced with ineffective corrective actions: it is considered sufficient to conduct a training course or publish some specific *vade mecum.*

It is like waiting to cure a bad sunburn with a good cosmetic moisturizing cream.

The problem to be solved is to **change business processes and individual habits that are not functional to a specific expected result**.

The first step, then, is to understand and share with the group what this "precise" expected result is and decide to achieve it:

■ Why do we want to improve the way we manage and conduct our meetings?
■ What do we want to achieve?
■ How can we get it?

If you want to halve the time you and your colleagues spend in meetings and at the same time find that you also bring more results, then you need to understand which habits are most useful—and how to make them take root—to achieve the result you want.

Understand the Value and Prepare to Get the Most Out of a Meeting

In Lean Thinking, any improvement process starts from the understanding of what constitutes value, before proceeding to the reduction and elimination of what is not, namely waste.

Why are we calling this meeting?

Investing time in the correct answer to this question can save us precious hours in all subsequent steps. The more precise we are at this stage, the higher the consequent benefits.

Clarifying the Purpose of the Meeting

A meeting can have several objectives and it is important to have a clear focus on each one, because this will guide the entire process of preparing and conducting the meeting in a Lean way. Here are some examples:

- seek information or suggestions from your group;
- provide information or updates to your group;
- involve the group in solving a problem or making a decision;
- clarify and define a problem well;
- share doubts or concerns with the group;
- share a problem involving people from different groups;
- solve a problem and make the relevant decisions, but it is not clear what the problem is or who is responsible.

Having a purpose for a meeting does not necessarily impose rigidity: a purpose could also be "exploring options," but the question you must ask yourself is always the same: what do you want to achieve with the exploration of options? You may want to make decisions, come up with ideas for improvement, understand who in your team thinks strategically. If you cannot clarify precisely what you want to achieve from the meeting, the risk is very high that the meeting will be unnecessarily long, will be attended by people who are not indispensable, or will go off topic.

- Write down a couple of quick notes on what you would like to achieve at the end of the meeting.
- Define the decisions you would like to make or the things you would like to say to help you focus and clarify.
- Evaluate if there is really a need for a meeting

Whatever you want to achieve through the meeting you want to do, the important thing is that it is formulated in such a way that it can be easily shared, communicated, and evaluated. At the end of the meeting, everyone should be able to say whether the expected result was achieved.

Clarify the Type of Meeting

One meeting is not the same as another: each one needs specific ways of preparation and management. If you do not understand well the difference between the various possible types, you will have difficulty understanding how to convey value and how to reduce waste meeting after meeting.

Therefore, once the purpose of the meeting has been established, the meeting must be framed according to its category. Below is a possible example of classification.

1. **Informative meetings.**

   If you want to convey or exchange information, provide, or receive updates regarding something new, the launch of some initiative, a change in the organization, processes, or the company in general, then this is the type of meeting to be organized.

2. **Consultative meetings.**

   They are meetings in which phenomena are analyzed, the causes of problems and their solutions are sought. A work plan is prepared, choosing who is responsible and the dates. There can also be meetings in which you seek information or suggestions from your group. Compared to the informative meeting, what changes is, therefore, the direction from which the information comes.

3. **Decision-making meetings.**
   These are meetings in which decisions are taken and choices made. Each choice requires at least two possible alternatives to choose from, so at least two proposals should be presented. In this type of meeting, it is essential, in addition to preparation, to be sure that the decision-makers are involved and that a follow-up plan is ensured once the decision is made.

4. **Relational Meetings.**
   They are a particular type of meeting where the focus is on the people involved in a specific ongoing work activity. They are usually done in a restricted form, sometimes in the form of one-to-one meetings, so much so that the very term "meeting" can often be considered improper, as meetings may even be held at the coffee machine or on a walk.

Defining the Agenda Well

The agenda represents the sequence of topics to be addressed and is one of the most powerful tools in meeting management. The agenda is indispensable to maintain the boundaries of discussion only on what is on the agenda and provides a good reason to stop any digressions on irrelevant topics or when you run out of time. It should be sent together with the invitation and is a useful tool to understand who should be involved and in which exact part of the meeting: if you are the organizer, always provide it; if you are a participant, always ask for it!

Define the timing in advance and assess the real feasibility of the various agenda items. Although it is not binding and basically represents a basic outline, in recurring meetings, such as stand-up production meetings or other alignment meetings, timing becomes indispensable. It helps us understand how far we are from what we expect, whether we need to cut some topics, speed up or, on the contrary, give more room for discussion.

Not every meeting needs a written agenda. For short sessions it may suffice in the first minutes of the conversation to agree on the agenda with the participants. Simply write down the main points and start the discussion.

Minimize the Number of Participants

If a meeting really must be held, only those who can effectively contribute to the expected outcome of the meeting must be included in the list of participants. There is no ideal number of participants, although many studies show that the efficiency of meetings is drastically reduced when there are more than 6/7 participants (Some studies have found that for each additional person over seven members in a group decision-making, the effectiveness of decisions is reduced by about 10%). It happens because, beyond this threshold of those present, who have something to say often do not have time to talk or to deepen, while others do not say anything and basically do only act of presence. The crowded meetings are often the realm of multitasking, where many spend time writing messages on their phones, practically without even listening to what they say.

Amazon CEO Jeff Bezos is known for instituting a "two-pizza rule": if you need more than two pizzas to feed everyone, the meeting is too big.

- Invite those who can benefit from the meeting.
- Invite those whose contribution to the purpose of the meeting justifies the attendance.

In a period of history that rewards agility and speed of action with respect to perfection, it is important that the pre-agility behaviors are spread in the company. The participation of the smallest possible number of people in each meeting is one of these behaviors. At the same time, to broaden the base of sharing and involvement, the contribution of those who do not attend the meeting can be requested in advance and meeting minutes can be sent to them. This way keeps non-participants in the flow of information and involvement, keeping them focused on their work and freeing them from the obligation of participation.

If a part of the meeting does not concern some participants, these people may not need to attend the whole meeting. If they are only needed for one agenda item, perhaps it can be the first item so that they are not waiting to be called to the meeting.

Can't just slim down the number of participants? A small exercise to start with is to eliminate the least necessary person from the meeting. This is not something your colleague has to take personally: explain to him that it is simply the best way to respect his precious time.

The Better a Meeting Is Prepared, the More Effective It Will Be

How Should Participants Prepare Themselves?

Should they bring data or collect specific information? Should they read something? Should they think about problem-solving? It is good to specify these kinds of requests when convening the meeting. It is important to let each participant know in advance what is needed to make the meeting as effective as possible. If you do not, you will spend much of the meeting doing unproductive group analysis and research on things that would have taken much less time before the meeting with individual work by some of the participants.

Have I Chosen the Most Appropriate Place and Tools for the Meeting?

In deciding the room, obviously depending on availability, distance from the participants and any cost, it is often appropriate to ask questions: is it wide enough (or is it too wide) to accommodate participants and equipment? Are the chairs comfortable (especially important for long meetings)? Or is it even better to do the meeting standing? Is there sufficient lighting and ventilation? Is it free from distractions, such as outside noise, telephones, neighboring rooms? Is it better to do it in-company or outside? Is it better to hold the meeting in presence, Covid-19 permitting, or at a distance? Is standard equipment, such as flip charts and markers, projectors, or monitors, available and functioning? Is water available to participants? If the meeting is long, is there some dried and fresh fruit to eat during the breaks?

What Is the Best Time to Hold the Meeting?

The best time is the one that suits the working routines of the meeting participants. A Lean Lifestyle Company should encourage, however, the routines that are most synergetic with the application of the operational principles analyzed in Chapters 5 and 7 regarding the simplification of working days. The first hours of the day should be left available for Gold Activities and for individual or small group work with high concentration. One to two hours before lunch and before the end of the working day are good windows of time for extended meetings. For production staff, at the beginning or after shifts may be the only opportunity to hold meetings without interrupting work. Borrowing an Anglo-Saxon terminology, we also talk about *brown bag meeting*, lunch or breakfast or aperitif meetings could also be considered: informative or relational meetings that can be held in an informal atmosphere inside or outside the company.

To realize recurring meetings, as we shall see later, the best way is to plan at the beginning of the year or semester, all the meetings planned for that period. When one does not attend pre-scheduled recurring meetings, the preferred alternative for scheduling a meeting is to schedule the next session shortly before the end of the current one.

Eliminate Unnecessary Meetings Is Better Than Making Them Efficient

Is the meeting necessary? Is there enough reason *for action*? We have already addressed this issue in the previous section, but it is worth emphasizing its importance before addressing the topic of reducing the duration of meetings. In practice, every time a meeting is convened, it would be necessary to understand if the topic can be addressed without bringing all participants to the discussion table, asking two questions:

What are the consequences of not having a traditional meeting?
How can I replace the meeting with other tools?

Many meetings can be replaced by a phone call, a two or three-way conversation, an e-mail, sharing documents, questions, or information on a collaborative platform. The aim is to open to the exploration of possible alternatives to meetings and to exploit the potential of ongoing digitalization. In a meeting, in presence or at a distance, communication is "synchronous," or live, in which everyone can listen and react in real time. If we want to eliminate one or more meetings, we must ask ourselves which ones can be replaced by "asynchronous" communication, typical of social platforms, where everyone can listen and read, but with possible reactions and interactions that can be deferred over time. If we do not see any negative consequences in making some meetings asynchronous, and if we even see benefits in comfortably reaching more people, in the times and in the ways that everyone prefers, we are faced with one of the fortunate cases in which corporate collaboration and social media platforms actually gain huge productivity and flexibility, expanding the frontiers of communication and engagement in an increasingly digitalized world. Some examples: Microsoft Teams, Slack, Facebook Workplace, Microsoft Yammer, Cisco Spark, Jive, Prysm Software Platform, Stride, Huddle, Salesforce Chatter. Most of these platforms offer the undeniable advantage of enabling synchronous remote communication and real-time collaboration as well as asynchronous forms of communication and deferred collaboration. The functionalities offered range from document repository and workflow management, the ability to create private or open groups for sharing files, information, questions and chat, collaboration tools including screen annotations, digital whiteboards, device sharing and support for web-based applications.

Eliminating meetings is possible, just learn how to do it, having the good sense to integrate traditional tools and new forms of digital collaboration.

Reducing the Duration Improves the Output of Meetings

We have arrived at the meeting. It must be done. There are no alternatives. Objectives clarified, agenda sent, preparation done. Whether it is done online or in person, we must set very clear objectives if we want to translate the Lean Lifestyle mindset here:

1. The meeting must be effective, in other words, it must be as effective as possible.
2. The meeting must be efficient, in other words, it must be achieved in the shortest possible time, with the least waste of time, energy and resources.
3. The meeting shall always seek to bring the maximum result with the lowest possible overall cost.

A good meeting starts on time and ends on time. Arriving on time at a meeting means showing respect for others and their agendas. Endless meetings always create frustration among participants, whose agendas go to hell without restraint. Whether it is a face-to-face meeting or a video meeting, you are not obliged to wait for latecomers and do not make the mistake of recapitulating what has been discussed so far, as you would risk wasting the time of all the other participants. Punctuality, however, is a necessary but not sufficient condition to ensure maximum effectiveness. Let's see how to act on the duration of a meeting.

Avoid River Meetings

Never spend more time than the bare necessary. How many breaks should you take during a meeting? None. Reduce the meeting duration to a maximum of 90 minutes and you won't need that break: meetings often take longer than necessary just because people can't get back from breaks on time. How many times have you seen everyone return at the exact time the break was supposed to end? Breaks almost always interrupt the flow of a meeting. Do you need more time than 90 minutes? Only in the very few cases where more time is needed, break the meeting into two or more parts and insert one or more breaks.

■ Experience regular quick "conversations" or standing meetings, as many companies are now starting to do, to reduce meeting durations. A 1999 study in the Journal of Applied Psychology found that traditional meetings with seated people last 35% longer than those held standing, with no difference in effectiveness.

■ To sensitize everyone to the habit of punctuality, conciseness, and time-keeping, start using highly visible signs for time management, such as large countdown displays. You will smile at first, as you do whenever you do something you have never done before, but then you will realize how useful it is.

PC and Mobile Policy

If you remember and share the principles of *selective silence, sacred time block* and sprint/relaxation, seen in Chapter 5, you now have the opportunity at corporate level to decline them in practice: prevent the use of PCs and phones during meetings. Stop. It is absurd to witness scenes like "I can't talk now, I'm in a meeting, can I call you back later . . . ?" It is absurd to see participants sending e-mails and WhatsApp messages while others talk. Every minute of time lost by each participant reflects on the effectiveness, emotional and mental climate, as well as leaving deep grooves in the actual culture of the entire group. Think about it. How do you feel when you see your colleagues or your boss or your managers, blatantly thinking about something else while you are talking? How do you feel when you realize that what you said and showed was not even heard or seen? What kind of culture are you experiencing in the company when in a meeting everyone is distracted and experiencing parallel moments with smartphones and laptops? Surely you are not living in the culture of maximum concentration, maximum effectiveness, minimum time, mutual respect. Isn't it better to strive to make meetings take much less time and find more high-quality time even for the necessary communication via e-mail or telephone? This is Lean Lifestyle put into practice.

Many companies that I have followed with the Lenovys team in recent years have introduced the concept of FAR WEST meetings, that is all inside, but the technological guns out. To trigger the new habit at each meeting, they use visible signals: brightly colored containers and clearly demarcated areas to deposit the technological tools before the meeting starts.

Lean Lifestyle Story: Farmalabor

The Meeting That Generates Value, for Oneself and for Co-Workers

250 hours saved each year. One and a half months of equivalent work dedicated to value-added activities for oneself and the company. More focused, less stressed and more results-oriented employees. The result was obtained thanks to a structured Lean Meeting project.

Francesco Ventola, general manager of Farmalabor, a company from Apulia specialized in the production and marketing of raw materials for pharmaceutical, cosmetic, and food use, describes his "encounter" with the Lean Lifestyle in the company as follows.

> Until two years ago, I did not control my time. I listened passively to requests and moments of confrontation with my 7 employees responsible for the operational areas of the company. We had not structured a regular meeting schedule and so when I did my One-to-One with them, the time spent was about 7 hours per week and the effectiveness of these meetings was low, mainly because there was a climate of continuous emergency.

Realizing the bad time spent on meetings, Ventola and his staff acted by setting all the appointments for the alignment meetings, the duration (15 minutes), the day and the time of day (after lunch) a year in advance. Each meeting has the shared objective of analyzing the department's production performance indicators and analyzing solutions to problems or proposals for improvement. If the meeting cannot be held in person it is still held, but on the phone. If there are no relevant updates, the meeting can be quickly closed or a discussion on issues of a personal nature can be held.

The signal to build and maintain overtime the habit was double: the appointment set in one's own calendar and a Visual Management paper tool behind the general manager's desk on which the appointments made during the year are marked for each employee. The gratification?

"Meetings always end with a coffee or in summer with an ice cream. Apart from the time regained for me, the most effective problem-solving, the impact on my well-being and that of my colleagues, has been enormous. It sounds paradoxical, but I am closer and more present than before: the less time spent in meetings has turned, thanks to a structured method, into more value for all of us and of course for the company.

Never Meetings without Decisions

Each meeting has its own rhythm marked by a sequence of operations that begin before the meeting, with the preparation stages, and are divided into several phases (Figure 9.1). The introduction serves to briefly warm up the engines; it should remind everyone of the agenda, the context, and the reason for the meeting. The purpose should clarify what you want to achieve and its end. The performance comes alive according to the procedures necessary for each type of meeting. The final stages are the most underestimated and taken for granted: in most cases there are hasty and hurried closures, given the frequent delays in closing meetings. The banana peel

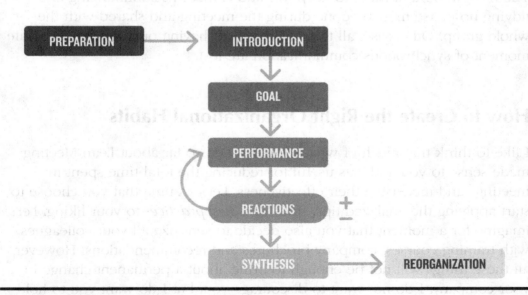

Figure 9.1   The standard process steps of a Lean Meeting.

on which many fall is there ready to make you fall and undo much of the work done because you lose the focus on what you want to happen after the meeting and often you even make the worst mistake of all: you finish without having made any decisions. And I am not talking about decisions on the highest systems, but decisions on who does what and when.

The moments of synthesis and reordering are crucial. Skipping this step means leaving space for other useless meetings to be held to pick up the discourse not concluded in the previous meeting, suffering other interruptions due to things not clarified at the end of the meeting, having to spend unnecessary time because of actions not done because no one had realized they had to be done, spending more time than necessary in preparing for the next meeting, and swearing because things do not happen.

The final rearrangement can be done by the speaker or the meeting facilitator, with this possible sequence:

- warn that closure is approaching;
- define the action plan;
- make a summary of achievements;
- describe the next steps: who-does-what-when;
- clarify the expected outcome of each action;
- fix the next follow-up meeting or milestone.

The most important thing to keep in mind is that this summarizing and tidying up phase must be done during the meeting and shared with the whole group. Otherwise, all the advantages of having organized a corporate moment of synchronous communication are lost.

How to Create the Right Organizational Habits

I like to think that much of what we have seen so far about Lean Meeting made sense to you and was useful for reducing the total time spent in meetings and increasing their effectiveness. Let's assume that you choose to start applying the analyzed tips, rules, and *best practices* to your liking. Let's imagine for a moment that you also decide to sensitize all your colleagues with training courses, company brochures and recommendations. However, all these things will not be enough to bring about a permanent change in your company. I do not want to discourage you, but I do want you to hold back your enthusiasm for the last step of the climb: learn how to design specific organizational habits and a corporate Lean Meeting system, which also includes systematic monitoring of deviations from the expected results. At each deviation you will have to stop a little, just enough to understand the reasons, to give and receive feedback, and do continuously evolve and calibrate the built system.

> The tools and arguments of Lean Lifestyle are applied and applicable in any company by anyone. This is the obvious difference I have found compared to the countless books and theories on the subject.
> **Federico Vannini,** Managing Director Sofidel UK

If we want to make a new behavior take root in the company, so that it becomes a new habit, we must necessarily follow certain criteria derived from scientific evidence matured over decades of study and analysis of human behavior. The first is to create the right antecedents or trigger the desired behavior with appropriate behavioral activating signals. Example from everyday life: at the red signal of a traffic light a person, not in the grip of alcohol or running to an emergency room, generally stops. The second criterion is to define concisely and concretely what behavior is to be followed, without the possibility of subjective interpretations. What to do in front of a red light is not interpretable. It is simple, clear, and unequivocal that you must stop! The third criterion to follow is the less obvious and less used in business: create relevant consequences, i.e. reward the desired behavior, and discourage undesired behavior. The consequences should aim at the consolidation of certain behaviors and the extinction of others. If a traffic policeman stops you because you ran a red light, or a camera caught your violation, there is only one consequence: a heavy fine and points reduction on your driving license. An intelligent way of setting consequences in the organizational sphere, which has been consolidated by a great deal of research on the subject, is to provide timely and incisive feedback to people based on observed behavior. Positive, quick, specific and concrete feedback will act as a rewarding consequence, reinforcing what has been observed, while negative feedback, provided in the form of neutral, non-judgmental observation, aimed at conscious reflection, will act as a disincentive or extinguishing consequence of unwanted behavior.

Verification of Shared Behavioral Rules

One way to implement a quick and effective feedback system on meeting behavior is to carry out a quick check at the end of the meeting to see what did or did not work. In Figure 9.3, you can see an example of a behavioral *hansei* activated to consolidate Lean Meeting behaviors in production meetings at Streparava. The verification at the end of the meeting really lasts a handful of minutes, but it provides precise, concrete, and quick feedback to the participants, simply providing input to foster awareness and reflection, without judgments of merit (Figure 9.2). If the group has actively collaborated in the construction of the list of target behaviors, these short follow-up sessions will be instrumental to the group's stated outcomes. Here are the steps to create Lean Meeting monitoring checklists:

■ identify key criteria and behavior with the working group;
■ define procedure for their monitoring;
■ defining the rewards linked to the achievement of results.

This is an example of a list of behaviors to choose from to introduce in the checklist:

■ Was the meeting well prepared?
■ Did the meeting start on schedule?
■ Have we completed the action plan before this meeting?
■ Was there active participation by all present?
■ Have we managed to keep the participants' attention and interest high?

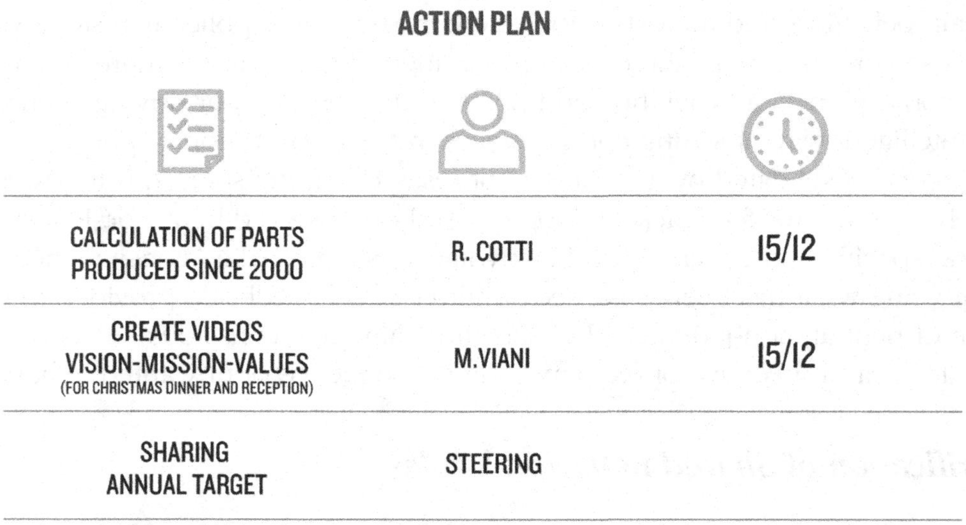

ACTION PLAN

CALCULATION OF PARTS PRODUCED SINCE 2000	R. COTTI	15/12
CREATE VIDEOS VISION-MISSION-VALUES (FOR CHRISTMAS DINNER AND RECEPTION)	M.VIANI	15/12
SHARING ANNUAL TARGET	STEERING	

Figure 9.2 Example of action plan at the end of the meeting.

- Have we listened and spoken one at a time, with moderate tones and appropriate vocabulary?
- Did the phones and PCs remain switched off, except for those who used them to project slides and data?
- Have we defined an action plan and set up the next follow-up meetings?
- Did the meeting end at the scheduled time?
- Did we keep the focus of the meeting, without digressions?
- Did the meeting achieve its objectives?
- Were the meeting support tools adequate?
- Have we made a final summary of what was said and decided?
- Did we send participants a report of what was defined during the meeting with concrete results obtained, even if negative?

Verification can be "analog," as in the case of Figure 9.3, or "digital," as in the case of Figure 9.4, where a purpose-built web-app was used for behavioral monitoring as part of the implementation of a Lean Meeting system.

streparava	WEEK DAY 29					WEEK DAY 28				
	17	18	19	20	21	10	11	12	13	14
Preparation of topics to be covered	✓	✓								
Punctuality of arrival	✗	✓								
Execution of planned actions	✗	✗								
Listening to others and dialogue one by one	✓	✓								
Transparency and respect	✓	✓								
Action plan at the end of the meeting	✓	✓								
Respect of time limit	✗	✓								
Ability to synthesize	✓	✓								

Figure 9.3 Example of an analog checklist for monitoring the implementation of a Lean Meeting system.

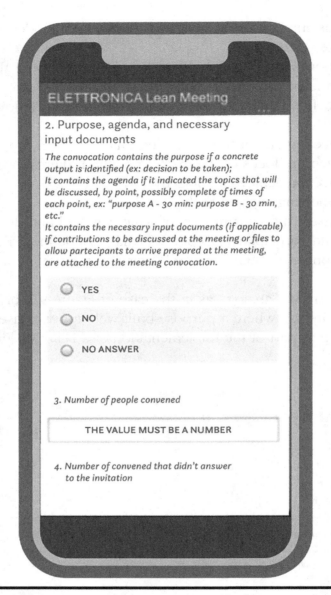

Figure 9.4   Example of a web-app for behavioral monitoring of a Lean Meeting project.

In the case of the digital solution, the data collected can be used for further analysis and the construction of more extensive and accurate metrics.

Set Business Standards

In behavioral analysis, a standard to be met should only be seen as an antecedent designed to promote the activation of the desired behavior. In "habit engineering" it is seen as the signal that stimulates the habit. In the words of Taiichi Ohno, as I have already reiterated in other sections of this book, where there is a process, there is a standard, and only where there is a standard, there is possibility of improvement. So if we want everyone in the company to convene a meeting, always indicating a clear purpose, an agenda, and necessary and optional participants; for example, it is much more effective to design a company-wide standard, share it with those who will use it most frequently, and incorporate it into common daily operational practice, until it becomes an established habit and, as such, automatic. Over time, the standards created can evolve along with the processes and new needs of the teams.

Examples of Standards That Can Become Lean Meeting Habits

What standards can we create to encourage Lean Meeting behaviors so that they become corporate habits? Here are a few:

- Document standards:
  - Standard agenda to be created for one-off and recurring meetings;
  - Meeting report.

- Standard for behavioral feedback:
  - Meeting preparation verification checklist;
  - Meeting management checklist.

- Environmental standards:
  - Introduction and check of working tools during face-to-face meetings (projection screens, connections, flipcharts, markers, etc.);
  - Visible stopwatches and countdowns for in-person meetings.

Standard does not always mean the same. They can also be constructed with interchangeable elements that "switch on or off" as required. As you can see in Figure 9.5, the agenda for this inter-functional monthly business management meeting is at the same time standard but variable month by month. If some urgency should arise, some of the scheduled elements can simply be switched on or off.

In Figure 9.6, you see an example of standards for the creation of a fixed agenda for a recurring meeting, enriched with other useful elements for the preparation.

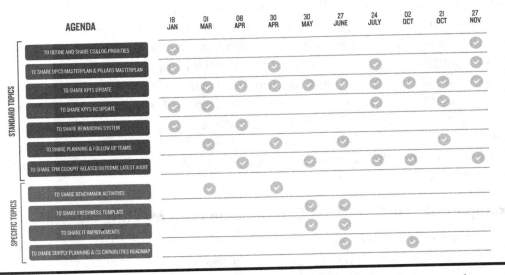

STANDARD DOESN'T MEAN "ALWAYS EQUAL"!

AGENDA	18 JAN	01 MAR	08 APR	30 APR	30 MAY	27 JUNE	24 JULY	02 OCT	21 OCT	27 NOV
STANDARD TOPICS										
TO DEFINE AND SHARE CS&LOG PRIORITIES	✓									✓
TO SHARE OPCO MASTERPLAN & PILLARS MASTERPLAN	✓			✓			✓			✓
TO SHARE KPI'S UPDATE		✓	✓	✓	✓	✓	✓	✓	✓	✓
TO SHARE KPI'S RC UPDATE	✓	✓					✓		✓	
TO SHARE REWARDING SYSTEM	✓		✓							
TO SHARE PLANNING & FOLLOW UP TEAMS		✓		✓		✓			✓	
TO SHARE TPM COCKPIT RELATED OUTCOME LATEST AUDIT			✓		✓		✓	✓		✓
SPECIFIC TOPICS										
TO SHARE BENCHMARK ACTIVITIES	✓			✓						
TO SHARE FRESHNESS TEMPLATE						✓	✓			
TO SHARE IT IMPROVEMENTS						✓	✓			
TO SHARE SUPPLY PLANNING & CS CAPABILITIES ROADMAP							✓	✓		

Figure 9.5   Sample agenda for inter-functional monthly business management meeting.

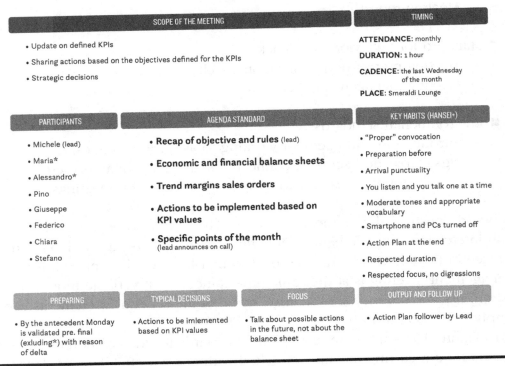

MANAGING BOARD

SCOPE OF THE MEETING
- Update on defined KPIs
- Sharing actions based on the objectives defined for the KPIs
- Strategic decisions

TIMING
ATTENDANCE: monthly
DURATION: 1 hour
CADENCE: the last Wednesday of the month
PLACE: Smeraldi Lounge

PARTICIPANTS
- Michele (lead)
- Maria*
- Alessandro*
- Pino
- Giuseppe
- Federico
- Chiara
- Stefano

AGENDA STANDARD
- **Recap of objective and rules** (lead)
- **Economic and financial balance sheets**
- **Trend margins sales orders**
- **Actions to be implemented based on KPI values**
- **Specific points of the month** (lead announces on call)

KEY HABITS (HANSEI+)
- "Proper" convocation
- Preparation before
- Arrival punctuality
- You listen and you talk one at a time
- Moderate tones and appropriate vocabulary
- Smartphone and PCs turned off
- Action Plan at the end
- Respected duration
- Respected focus, no digressions

PREPARING
- By the antecedent Monday is validated pre. final (exluding*) with reason of delta

TYPICAL DECISIONS
- Actions to be imlemented based on KPI values

FOCUS
- Talk about possible actions in the future, not about the balance sheet

OUTPUT AND FOLLOW UP
- Action Plan follower by Lead

Figure 9.6   Example of a fixed agenda standard for a recurring meeting.

Lean Lifestyle Story: Deoflor

Preside over and Lead the Company in Less Time and with More Results

How to reduce daily meeting time from 70% to 40%? This is what **Gian Luca Guenzi**, CEO of Deoflor, the only Italian company specializing in the contract production of Toilet & Air Care products, did. The systematic introduction of Lean Meeting and slotting techniques has literally changed the entrepreneur's way of working: "I used to work to make meetings and not to obtain results on activities of value for me and the company." Here are the structured actions assumed as work routines:

- Strict division of the day: morning dedicated to interaction with co-workers, internal and customer meetings; afternoon dedicated to focused operational activities.
- Schedule of alignment meetings set in advance with all co-workers:
  - Sales area meeting: 3 appointments per week with the manager and 3 meetings with the marketing and customer care area. Fixed duration of 30 minutes.
  - Board meeting: one 60-minute meeting per week.
  - Meeting with R&D: 3 appointments per week lasting 30 minutes.
  - Production visit: daily at 11am.
  - Meeting with industrial management: 3 appointments per week lasting 30 minutes.

This way of working and leading the company has served me well because I have a control of the whole company that cannot be compared to before. All areas have the right amount of time and focus.

Today, I can afford to handle extraordinary situations with a customer, which take a lot of time, without losing control of the rest of the business. The discipline and the solidity of the working method gave me and the organization simplicity.

Company Meeting System

Whenever you can, try to translate a recurring meeting into a systematic meeting: it brings efficiency, it generates a habit, people already know what they are going to talk about, what data to come up with, there will be no confusion about times or where the meeting will take place. As already described in Chapter 7, the annual calendar of different company organizational routines is an excellent exercise in reducing waste and interruptions in the preparation of different meetings. Making an annual calendar forces you to clarify how you want to govern the entire company from top to bottom.

This will be a topic in the chapter on the strategic management of a company.

Meetings are simply a tool. The objective of each company is to achieve its strategic and tactical goals: the various meetings should serve to make decisions, solve problems, monitor, and supervise at various levels what is happening toward the set direction. Here is a possible list of periodic meetings that can be part of the company's Meeting System and, therefore, of the annual calendar of the whole organization:

- Daily/Weekly management meeting;
- Cross-functional meetings on corporate governance;
- Strategy deployment meetings;
- Corporate communication and alignment meetings;
- Meetings on repetitive key processes;
- Periodic one-to-one meetings.

- **Daily/Weekly management meetings**
  They are short meetings at the most operational level of the organization. Every group in the company has its targets to achieve and its monitoring of these should be as frequent as possible: per working shift, per day, per week at the most. No more than that. What are the reasons that have prevented us from reaching our targets? What have we already done, what can we still do, what do we need, and who will do what when? What are the priorities of the day? Or the shift? Or the week?

- **Cross-functional meetings on corporate governance**
  These are periodic meetings to monitor and oversee both the current business and the strategic initiatives undertaken. The composition of the group of participants in these meetings may reflect the company's stable organizational characteristics or may involve different functions depending on the "piece of business" to be overseen. The projects to be monitored or the parts of the business to be overseen will determine who is required to participate in this type of meeting.

■ **Strategy Deployment meetings**

These are all the meetings a company must have each year to generate, develop and monitor the company's annual strategic plan. This topic will be explored in Chapter 13, where we will see the various steps required to create a company strategic plan consistent with the Lean Lifestyle strategy.

■ **Corporate communication and alignment meetings**

These are "institutional" meetings where the company periodically communicates context, challenges, goals, perspectives, and ongoing initiatives to everyone. They serve to keep everyone on board, celebrate victories and challenges won, and gather feedback from people. Depending on the size of the company, they can be collective or divided into appropriate portions of the company.

■ **Meetings on repetitive key processes**

These are meetings related to processes that must be conducted every year in the company, for example on budget, generating innovation ideas, performance evaluation, supplier days, customer days, etc. Taking care of these meetings in advance will enable all participants to prepare well, not to improvise and not to miss pieces along the way.

■ **Periodic one-to-one meetings**

We talk about periodic individual meetings for the development of your employees, to monitor their performance, delegate and supervise, and above all to give and receive appropriate feedback, whether positive or negative. Like a coach with his athletes, these meetings are far more effective than traditional end-of-year performance evaluation meetings. They are short, close meetings, where we can "straighten out" our pitch, without waiting too long, when we have little space to recover.

Perhaps all these meetings to be made repetitive and standardized may frighten you, but the reflection to be made is that all these activities, in most cases, are ones you already do or would like to do. Putting things in order and creating your own Meeting System, as in the example in Figure 9.7, will give you the opportunity to critically analyze how your organization is functioning, with the aim of making it smoother and leaner, ticking like a Swiss watch with the right cadences to achieve the desired results.

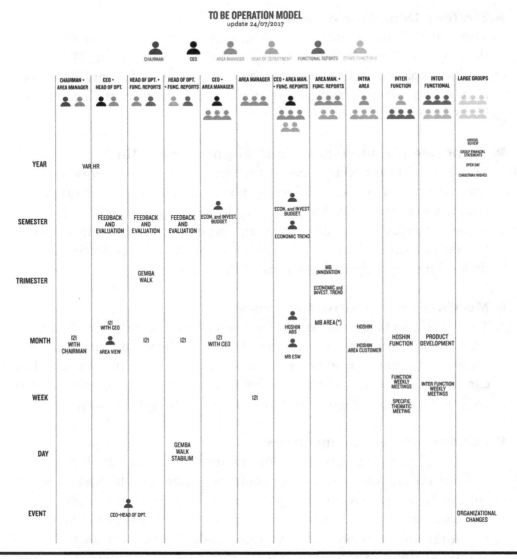

Figure 9.7   Design Example of a corporate Meeting System.

How to Consolidate Results and Improve Them over Time

"You only improve what you measure." I know. I'm repetitive. What metrics can be used to make sure that what we have seen in this chapter is having a real impact on your company? What steps should be taken to improve our Lean Meeting system over time?

Generally, the rationalization of the company meeting management system is carried out with a project involving one or more parts of the organization.

The initial situation and the progress indicated must always be quantified with objective measures. Here is an example of indicators that can be adopted:

- Number of meetings per day;
- Effective average duration of meetings;
- Number of people convened for meetings;
- Number of people present per meetings;
- Number of "useless" people per meeting (criterion: they did not contribute information and ideas for decisions);
- Number of overruns divided by the number of meetings;
- Planner ability indicator: number of "extraordinary" meetings divided by number of meetings held;
- Time lost due to delay: number of persons multiplied by number of minutes of delay;
- Number of minutes delay from the beginning;
- Number of minutes of overruns since start;
- Time lost due to overruns: *number of people multiplied by number of minutes overrun;*
- Meeting cost: number of participants multiplied by duration multiplied by cost per hour;
- Useless costs per meeting: (no. of useless people multiplied by meeting duration + number of minutes delayed + meetings with purpose achieved) multiplied by average cost per hour multiplied by number of participants;

Don't choose too many. Just a few are enough. Even one is enough. What matters is to have a sense of tangible progress. Also, because generally Lean Meeting behaviors act with a strange domino effect: I start with the habit of starting on time and inevitably end up being "obsessed" with duration and finishing on time. I start by limiting the number of participants and end up wondering each time whether we really need that meeting or not.

A useful expedient to do end-of-meeting verification and data collection for the measurement system at the same time is to combine the two needs, as you can see in Figure 9.8, where the Streparava verification checklist was also designed to be used as data collection.

Technological potential now makes it possible to build digital tracking systems that simplify collaborative flows, data collection, and monitoring. Asana, for example, makes it possible to create digital meeting agendas and

CONVENING

streparava.

1. Meeting type: ☐ EMERGENCY ☐ PERIODICAL ☐ OPERATIONAL

2. Number of **people convened** _____
   (with the exception of optional participants)

3. Number of **present members** _____

4. Does the **convocation contain** the **purpose** of the meeting? ☐ YES ☐ NO

5. Does the **convocation contain** the **scheduled agenda**? ☐ YES ☐ NO

HOST

6. Does the host let discuss only about the arguments
   scheduled in the **agenda**? ☐ YES ☐ NO

7. Does the host achieve the main purpose of the meeting? ☐ YES ☐ NO

8. Does the host summarize and write and share
   the **action plan**, identifying responsibilities and deadlines? ☐ YES ☐ NO

9. Does the host **respect the duration** of the meeting? ☐ YES ☐ NO

PARTICIPANTS

10. How many participants did not notify in advance
    of **presence** or **absence**? No. _____

11. How many participants **did not arrrive on time**? No. _____
    (including the host)

12. How many participants **did not mute** the smartphone
    and used the PC? No. _____
    (unless they had to)

13. How many participants **interrupt** who is speaking? No. _____
    (also for meeting reasons)

14. How many participants make interventions and/or
    inconsistent questions with the agenda points? No. _____

15. Is the meeting **interrupted from outside**?

16. Is there all the **necessary information**? ☐ YES ☐ NO

Figure 9.8   Example of a verification checklist used in Streparava.

to manage the action plan that is constantly shared with the whole team, monitoring in real time the closure of the actions resulting from the meeting.

In Figure 9.9, you can see a "scoreboard" of a company Lean Meeting initiative, made available in analog-paper and digital format. This display

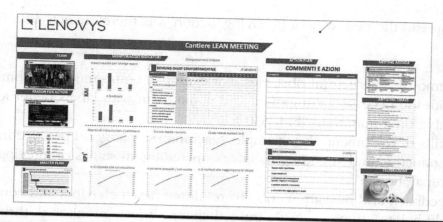

Figure 9.9 **Lean Meeting project board, with sharing activities, starting data and targets, efficiency indicators, checklists, and celebrations.**

board collects and makes usable the various progress of the working group: work plan, starting data and target goals, efficiency indicators, behavioral checklist and number of measurements taken, other communication, and celebration elements.

Lean Lifestyle Story: ELT Group

Results and Less Waste with the Lean Meeting

The bet: I will serve croissant and cappuccino every morning for a week to all the participants at the construction site. The challenge—to regain efficiency and effectiveness in meeting management—launched at ELT Group, a company in the defense sector, was so ambitious that the human resources director, sponsor of the initiative, jokingly put up his cappuccino and brioches delivery service as a "prize" for the 30 people involved in the pilot project.

The initial assessment was clear: only 10% of people were satisfied with the meetings because their preparation and conduct were riddled with waste such as the lack of key people, technical tools, shared agendas, and action plans at the end of the meetings. Above all, they had become aware that the cost of a meeting averaged €1,000 and that the cost of initial delays alone to "wait for latecomers" amounted to €100 per meeting.

The key steps: removal of obstacles, training, monitoring, and feedback.

After the initial assessment, the project team identified the behaviors that made the meetings inefficient and then the countermeasures, the key behaviors to achieve improvements.

To spread the new way of behaving, action was therefore acted on three fronts:

- Removal of structural obstacles. For example, equip meeting rooms with the necessary equipment and name them with labels better identifiable on a map than "ELT2530."
- Skills development. For example, teaching people good practice in conducting and attending meetings, including all stages of meeting preparation.
- Feedback and measurements. To reinforce new behaviors and combat "old habits," some specially trained change agents monitored, throw an anonymous checklist, the behaviors of participants in the meetings in which they were involved and provided timely feedback at the end of the meetings to colleagues. In addition, at a monthly meeting in the presence of all project participants, indicators were discussed, and improvement initiatives were taken, for example, the decision was made to install a clock in the meeting rooms to help with timekeeping. This led, in the pilot project alone, to a total of 165 completed checklists and 700 returned feedback, mainly positive, but also corrective.

The Director Served Cappuccino! The Results of the Pilot Project and Its Extension

At the end of the project, the punctuality of meetings in the pilot area increased from 59% to 72%; the purpose of meetings was achieved in 96% of cases, whereas previously it was 48% and so the duration of meetings reached 96% (starting from 48%).

These are some of the indicators that made the project team toast to success and led, with great satisfaction, the personnel manager to make his "penance."

After this first milestone, the project has stepped up and started its expansion phase to the rest of the company, involving around 500 people, with a monitoring of over 1400 meetings, including online meetings that started to spread after the Covid-19 "lockdown."

The monitoring system has been evolved and digitized so that the checklists can be filled in from smartphones and graphs with trends are automatically available in real time.

Through the same action scheme of the pilot project, suitably adapted to make it sustainable on a large scale, equally positive results were achieved in the expansion phase.

Despite the spread of chronic behaviors in the company that were feared would be difficult to eradicate, meetings with a clear agenda and goals have increased from 76 to 90% of cases; the reception of a response to the convocations by all the participants increased from 74% to 89%; the punctuality of the beginning of the meeting (a real "black beast") from 63% has approached more and more to the target of 85%; finally the meetings in which only the agenda is discussed without digressions reached 95%.

"Beyond the objective and measured results, I can say that a project started with a lot of skepticism from those who did not know the method, and from a part of the corporate population less inclined to change, has generated important and lasting effects not only in the population involved," comments **Giulio Cesare Grande**, Engineering Manager of the EW RADAR & EWCC Product Lines and responsible for the Lean Meeting project in ELT Group— Certain behaviors were also contagious even in departments that were not involved in this initiative. Prior to this "intervention," there were mainly one- or two-hour meetings, many of which had no agenda and therefore it was not clear how the meeting duration was decided; the late start was endemic and widely accepted; the end of the meeting often exceeded the schedule. Now the situation is very different: all the meeting rooms have been renovated and structural obstacles removed; they all have monitors or projectors and flip charts; they are bookable from Outlook and have a screen indicating who booked it and the duration. If you do not confirm your booking on the screen with your pin after five minutes from the start, the room is automatically vacated. This is to discourage delays. Now those who arrive late apologize and delays are very limited. There are even meetings of 30 or 45 minutes, and most meetings finish at the scheduled time.

Giulio Cesare Grande points out, finally, that the introduction of meeting management technology came about thanks to the work carried out to measure (and then objectivize waste and the related costs) and how the overall results have gone

beyond expectations, and this can be seen in the way meetings are now managed in the company, even by new recruits and those who did not take part in the project: the proof of how the behavior introduced has become corporate culture.

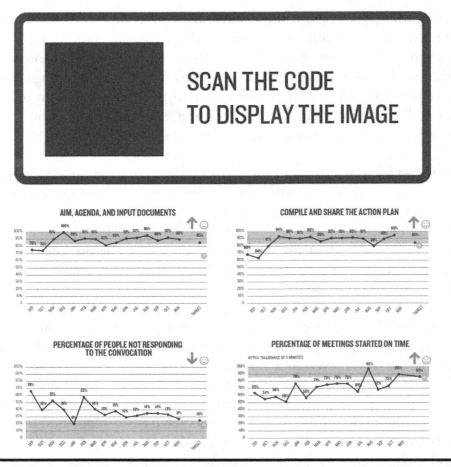

Figure 9.10   Example of real-time KPIs monitoring system on some of the key Lean Meeting behaviors.

Lean Lifestyle Story: Fire

War on Disorganized Meetings

The war against the ineffective management of meetings at Fire spa—Italy's first independent player in credit support services—has seen its CEO Claudio Manetti as a proud leader. He decided to start right from the management meeting with his front line: typically, very messy meetings that had to last 3 hours and regularly "overran" abundantly.

The activities implemented were manifold. In the first instance, it has raised awareness of the problem with its staff, by explaining to them the effects of working in multitasking mode.

Then all together they defined and shared the new rules for the preparation and management of each of those meetings:

- creation of a standard agenda;
- limits on the maximum number of slides in presentations;
- classification of interventions into "informative" and "decision-making" to improve the listener's focus;
- Telephones outside the room or switched off;
- drafting, during the meeting itself, an essential record with key actions and decisions;
- use of a large clock in the center of the table as a signal for punctuality of interventions;
- regulation of applications, by reservation or in a dedicated slot at the end of the presentation.

Thanks to these rules, the staff meeting has become much more effective and has been reduced from 3 hours regularly overrun to 2 hours. Next goal: to extend this approach to all meetings in the company.

Says **Claudio Manetti**:

I put myself on the line personally, because I had to be the first one to set a good example. We have also tried to introduce new behaviors using a playful system of scores, rankings and consequent teasing based on whether presentations were completed early or late. In addition, we helped each other abide by the rules by playing around and giving feedback to each other. We have also got used to the logic of experimentation: at the end of each meeting, we have 2 minutes of hansei on how the meeting went; if we find an improvement, we try it out and if it works, we change our approach, in full agile spirit.

Lean Lifestyle Story: Streparava

The Power of Feedback

Meeting management at Streparava is now a flagship and a reason for deep satisfaction, looking at the starting level. "We were a disaster," says **Raffaella Bianchi**, Human Capital Manager of the Brescia-based company, a reference player in the automotive sector for the development and production of Powertrain and Driveline systems.

It was a constant delay of the participants or the presence of people who did not know why they had been summoned. We tackled the problem by "deploying" 20 colleagues in charge of compiling a checklist during the meetings containing misbehavior: with this information we drew up a decalogue and provided specific training on the Lean Meeting. Shared the route, we started to run the meetings according to the rules we had set ourselves and at the same time started the punctual monitoring of all the meetings by the figure invested with the role of "chairman," in charge of providing positive and negative feedback to participants.

The approach has proved to be successful and meetings management has become more and more refined. Raffaella explains again:

Swiss punctuality, in presence and online. Shared agendas, preparatory documents available to all participants, action plans always compiled at the end of the meeting, it has become normal to prepare and manage a meeting to be productive and not a waste of time. In particular, it was crucial to identify the figures of the meeting chairmen: Their role above the parties makes them responsible for monitoring and feedback: they can "chastise" even a superior and this is not only accepted, but has fostered a domino effect that facilitates self-control and the attention of everyone to respect our decalogue (Figures 9.11 and 9.12).

Entering a meeting in Streparava one is certain not to find smartphones switched on and, above all, consulted. Adds amused Raffaella:

The consequence? The switchboard is flooded with phone calls! Before, those who looked for us with the direct number were almost certain to find an immediate answer, even in a meeting. Not now. But we have found a solution: we are accustomed to those who want to talk on the phone, to write an e-mail, to schedule an appointment, and for insiders we encourage the use of the internal chat where they can leave their request. In this way we achieve the double result of not being interrupted and slotting for phone calls!

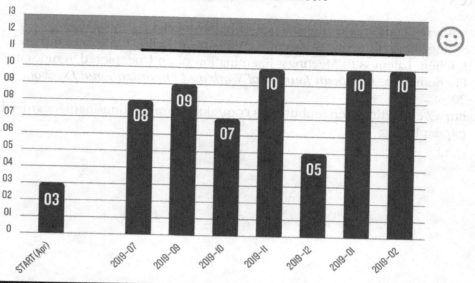

Figure 9.11 **Results of the monitoring of the application of the behaviors of the Lean Meeting project in Streparava.**

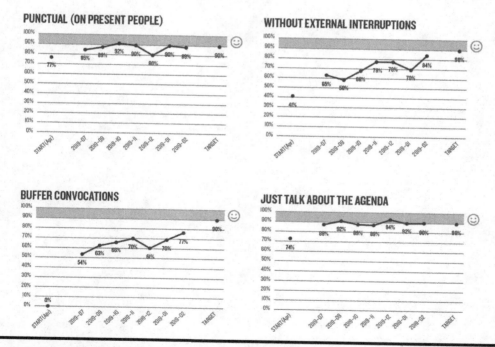

Figure 9.12 **Extract from the Streparava system of real-time KPI monitoring system on some of the key Lean Meeting behaviors.**

Notes

1. www.wsj.com/articles/the-science-of-better-meetings-11550246239
2. S. Rogelberg, C. Scott, B. Agypt, J. Williams, J. Kello, T. McCausland and J. Olien, Lateness to Meetings: Examination of an Unexplored Temporal Phenomenon, *European Journal of Work and Organizational Psychology*, 2013.
3. https://e-meetings.verizonbusiness.com/global/en/meetingsinamerica/uswhite-paper.php

Chapter 10

The Five Levers to Explode the Value of People in the Company

The ATRED Method

An improvement project is not denied to anyone! It is becoming more and more common to find companies resorting to initiatives, structured, with the aim of increasing a performance: for example, plans to improve performance and reduce the costs of their products, office processes, production processes, conversion rates of sales offers or contacts arriving from the web.

There are those who make improvement projects based only on their own strengths and knowledge, sometimes appointing dedicated internal figures for the precise purpose of pursuing "continuous improvement," or there are those who rely on external consultants or companies. Finally, there are those who do none of these things, but here another chapter opens, namely that of companies and entrepreneurs who entrust their destiny to fate. And unfortunately, this is the same story for all those people, in the company and outside, who do not bother to elaborate and implement structured individual improvement plans to achieve their excellence and express their full potential. In companies, there is also a widespread superficial attitude toward everything that concerns improving the living and working conditions of the most valuable resource: people.

Apparently, today, entrepreneurs and managers agree that the personal and professional well-being of people in the company is an important

DOI: 10.4324/9781003474852-10

element of success and growth. Many like to publicly define people as their "main asset" and boast of putting "people at the center." However, **there are many who underestimate the real impact on business and the real welfare-results correlations**. Few make operational tools available in a systematic way for this purpose.

Whenever people's well-being increases, both their individual performance and the company's overall service level and profitability increase.

Many studies agree on this direct correlation. Among the most well-known, for example, is the one published by the Harvard Business Review, according to which the well-being of people leads to a 31% increase in corporate productivity, 37% increase in sales, and a threefold increase in creativity levels.[1]

Yet, we enter the company every single day of our working life and start "doing things," often complaining that we never have enough time and energy to do this or that thing we have been procrastinating on for months and perhaps years. We accept a sub-optimal level of well-being and a level of performance that does not correlate with our efforts.

We never stop to reflect on how we can free up time, review our priorities, work better, live better, achieve more with less effort and more well-being. Unfortunately, we fall into the silliest, but at the same time most common mistake, of devoting equivalent amounts of time to activities of vastly different impact. For example, one hour to absent-mindedly read e-mails or wander the web, and one hour for a project that could radically change the fate of the current year. And I have been magnanimous with the proportions. It often happens that the amounts of time are not quite so salomonically equivalent but instead are mercilessly shifted in favor of the activities with the least impact!

We have already addressed this topic in Chapter 3, High-Impact Work, where the importance of Gold Activities in our personal and professional lives has been discussed in depth. We analyzed in depth in Chapter 4 all the performance and health consequences of a fragmented and disordered way of working, characterized by excessive multitasking and very frequent changes of context.

We have analyzed in the following chapters many possible individual and organizational countermeasures to simplify and fluidify the working days.

What now?

"People are insane because they want to have different results by continuing to do the same actions as always," Einstein warned.

Something different must be done. But it is necessary to have a method that can affect the personal and corporate working methods and, at the same time, to trigger positive tangible impacts at the individual and organizational level.

In this chapter, we will analyze a tool that summarizes this method, to begin to concretely eradicate old habits in favor of new ones, increasing people's performance and well-being. It is a tool that is conceptually simple, but at the same time powerful in terms of the results it brings.

This tool works under two conditions.

■ The individual must accept that he or she is entering the personal highway of continuous improvement, where every single meter traveled moves you forward and progresses toward the desired goals. And as in any motorway journey, you must first want to reach your destination, accept the fact that you cannot jump from where you are now to where you want to go: the only chance you have is to get moving and check the progress kilometer by kilometer.
■ The company must accept its role as a supporter and promoter of the individual improvement of people, spreading the tool as a routine practice of its organization.

The Five Levers of the ATRED

Archimedes has been attributed one of the most beautiful motivational phrases I know: "give me a foothold and I will lift the world."

On that point, the good Archimedes would have placed a lever to lift anything. And it is precisely the concept of a lever that I used to develop this method with the Lenovys team. We called it **ATRED,** an acronym made up of five letters, each of which represents a lever for realizing individual time, space, and energy:

■ Ability (A): competence lever.
■ Time (T): Own time lever.
■ Reduce (R): efficiency lever.
■ Eliminate (E): radical elimination lever.
■ Delegating (D): lever of the time of others.

1) **Ability (A): competence lever.** Already mentioned as one of the key factors in Gold Activity, it is a very important lever to use to acquire a "superpower" in any sphere of action. The more skilled

and competent I am at doing a given task, the less time it will take me and the more effective I will be at completing the tasks.

2) **Time (T): own time lever.** You will not detach an inch from your current situation unless you decide to balance your time differently. You cannot keep doing the same things, taking the same amounts of time, and getting different results. You have to decide which activity, among the many that you do, deserves significantly more time than you do today. The time lever may seem trivial to you, but it is the one that will make the most difference on your journey to personal excellence. And it represents the main challenge to be won. A challenge, however, that is impossible to overcome without the combined use of the next three levers.

3) **Reduce (R): efficiency lever.** If I decide to add time to something, you will have to reduce it accordingly elsewhere. It is nowhere written that everything you do must take exactly the amount of time that you commit today. When you start using this lever, you will begin to ask ourselves how we can reduce the duration of each activity by looking at them with critical eyes. Then you will find solutions that will show you that we will never again have to assume that the time you need corresponds exactly to the time we have been using up to that point in the various activities, partly out of habit and partly because you end up becoming attached to the things you do and the way you do them.

4) **Eliminate (E): radical elimination lever.** This is the lever that gives the most adrenaline rush (to me and to everyone who has tried it!). Deciding to eliminate activities *tout court*, overcoming embarrassment, fears, and indecision, gives you a courageous leap forward because you immediately gain the entire portion of time gained. A few examples? Recurring events or meetings that you attend and that you decide not to attend any more; commitments that you decide to cut below a certain critical threshold, such as sales visits to clients below a certain business potential; reading reports, articles, journals, dossiers, and analyses that you no longer consider relevant to you; an occupation that you no longer enjoy. And so on. Obviously, you will have to decide how to handle what you have eliminated. You may already be thinking of delegating it, but what I suggest you look for, in applying this leverage, is pure elimination. Without delegation. At the most, you will have to try to build a "system," automatic or not, where you no longer need to be

present to manage the activities that you do not want to completely abandon, despite their minor impact on you.

5) **Delegate (D): lever of others' time**. We have seen in detail the "superpowers" of high-impact delegation and how little it is still fully exploited by many entrepreneurs and managers. Many people find it difficult to delegate because they are either intimately convinced that they are the only ones who can perform tasks better than anyone else, or they are equally convinced that they have no people around them to whom they can delegate anything. Let's think about it for a moment: if by some absurdity the value of my work were twice as great as that produced by others, it would never be equal to what three, five, ten or a hundred people can produce. Just as it is absurd to think that with hundreds of people directly or indirectly connected to us, there is no one to whom we can effectively delegate some of our activities. To use this leverage well, therefore, we need to break down some heavy limiting beliefs. Human beings, in general, do not like to delegate because they do not like to lose control of the situation. The feeling of control makes us feel stronger, gives us more security, more power. And when we delegate, we have the initial impression of losing some of this power, leaving us almost at the mercy of others. When we replace the word "control" with the word "presidio," we have already seen it, and we can in fact open endless development opportunities for us and for others.

Steps for Processing Your ATRED

In companies of any size, the ATRED method can be applied individually or be used to guide employees on a path of individual improvement. Let's see how.

The initial goal is to map all the individual activities we are involved in, classify them in order of relevance, and quantify their duration. In this way, we will have an overall representation of the time spent and the various items that are part of it in detail.

The ultimate goal is to identify and implement concrete actions to reduce the number of committed hours and rebalance everything to go with greater security and in the desired direction.

These, in summary, are the steps to be taken:

1. List of activities carried out and assign the total duration to each one.
2. Identify Gold Activities.
3. Create your own improvement plan.
4. Monitoring the implementation of actions.
5. Periodically review the ATRED

If you want to try the ATRED method now, you can print the format in Figure 10.1 downloadable from my site and get to work.

Try to include in the relevant column all the activities that come to mind, especially the recurring ones, and avoid spot activities or activities that you only do a few times a year**, I strongly advise you to time the actual durations of activities for about two weeks**, because subjective perception almost always misleads us. Just as I advise you to objectively measure the number of interruptions during the day. Enter all the hours worked, including those done at home so that the total number of hours is the actual number of hours worked, for example reading e-mails from your mobile phone at home and working at the weekend or in the evening.

	ACTIVITY	ACTUAL TIME HOURS/WEEK (hh:mm)	IMPACT	PASSION	ABILITY	CHALLENGE	GOLD	A ABILITY	T TIME	R REDUCE	E ELIMINATE	D DELEGATE	TARGET TIME HOURS/WEEK (hh:mm)	+/-	IMPROVEMENT IDEAS
01.															
02.															
03.															
04.															
05.															
06.															
07.															
08.															
09.															
10.															
11.															
12.															
13.															
14.															
15.															
16.															
17.															
18.															
19.															
20.															
21.															
22.															
23.															

Figure 10.1 Downloadable ATRED format.

For the definition of the gold activity, remember to use the eight-question route we have already examined in Chapter 3: the Gold Activity Scouting. If you have not yet done so, you can take the chance to combine the two tasks into one. By processing the ATRED, you will add quantitative elements to the Scouting that you did not use before: hours spent and frequency, current and future.

The four key factors, impact, skill, passion, and challenge, will be attributed to each activity to make it easier to recognize your Gold Activity in a single working tool.

1. **Impact**: ask yourself what the impact of each activity is compared to another to identify the non-value-added activities that act as ballast.
2. **Abilities**: try to understand which activities you are more skilled at than others. This will help you not to burden yourself with things you are not good at, to be aware of your strengths to make areas of excellence and to decide how to deal with the weaknesses you have identified: eliminate, reduce, delegate, or transform into strengths.
3. **Passion**: try to understand which activities excite you. An activity will never be Gold if you don't enjoy it; it is difficult to achieve excellent results doing things that you neither enjoy nor stimulate.
4. **Challenge**: Don't stay in your comfort zone; identify activities- current or new to add- in which you can experience the concept of challenge for your continued growth and to overcome daily limitations.

There are no automatisms: you choose your Gold Activities at the end of the day. Assessments of key factors should only guide your thinking. It might be very useful to talk to your colleagues or your manager to get their views on the Gold Activities related to your organizational role.

Choose, then, your Gold Activities among those that you think could take you to a goal synonymous with excellence for you. Choose them from what you do today, perhaps devoting more time to it, or from what you do not do and could do to reach a goal that is significant to you.

Finally, use ATRED to create your improvement plan using the 5 levers:

- **A** (Abilities) = Example: training to increase skills.
- **T** (Time) = Example: increase time for Gold Activities such as staff development or strategic activities.
- **R** (Reduce) = Example: eliminate waste, improve concentration to be more productive.

- **E** (Eliminate) = Example: eliminate useful tasks.
- **D** (Delegate) = Example: decide what to delegate to make room for Gold Activity or new activities

Define the levers you have chosen to use and elaborate on the improvement actions you intend to take concretely in the next three months. Don't be vague: the more precise and specific you are, the greater your chances of success will be. Complete the work by including the estimated targets for improvement. The effect of the actions you have decided will be threefold:

- Free up time thanks to the reduction of waste
- Increase the time devoted to Gold Activities to get more results
- Growth of well-being by implementing specific improvements in working conditions and consciously introducing activities that simultaneously increase impact, satisfaction, and passion.

In our business consulting projects, we help people, step by step, in the correct setting of their ATRED, as well as improve actions and provide support in subsequent periodic monitoring. To understand whether one is going in the right direction, we use graphs like the one in Figure 10.2 to

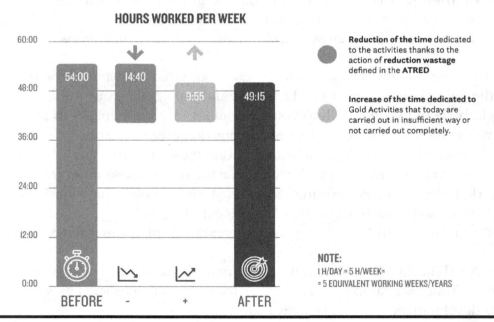

Figure 10.2 Example ATRED results graph.

quantify how many hours per week have been gained and how many hours have been converted into Gold Activity. After almost ten years of using this fabulous tool, I can say that it is difficult to find cases where less than two hours a day are earned.

A key habit that makes the difference between those who get results and those who don't is to set a sacred time slot each week in which to work on your ATRED and reflect on how improvements are progressing.

Reduce, Eliminate, and Delegate: The Secret Weapons of the Explorers of Lost Time

Throughout my professional life and with the experience I have had with my team in hundreds of companies, alongside thousands of people, the first reaction to the suggestion of reductions or deletions of activities is as follows: "Oh well, now they come along and tell us that we can reduce or eliminate something that we do every day. It's not like we're idiots doing useless things . . . ! And what do they know about what we do! What we really need is someone more to help us do the many things we have to do . . . ."

After more than 25 years, I can say that to solve critical situations, in terms of saturation and overload of commitments, it is never enough to add more people to the teams. On the contrary, I have often seen worse. What can improve overloaded situations is the choice to mercilessly cut activities and projects. What happens to a garden when it is left to itself, without regular pruning and maintenance? Alongside the trees and plants that were initially chosen, weeds and weeds will grow, leaves and debris will accumulate, perhaps unharvested fruit will go to waste, and most likely even the initial plants will have grown out of all proportion, occupying more space than desired. The same thing happens with everything we do every day. Commitments pile up, urgently overlap with each other, and we really struggle to distinguish what we can reduce or eliminate from what we necessarily have to do. Just as I mentioned pruning and recurring maintenance in the garden metaphor, in the same way, in my opinion we should introduce periodic pruning and routine maintenance of our "set" of acquired commitments. The ATRED tool facilitates us in this operation because through a systematic overview, it solves the common mistake of focusing on partial aspects and commitments that do not help in deciding what to keep and what to let go. Let me give an example. When I focus only on one activity, isolated from the others, it always seems feasible to me to sustain that commitment, because I am not aware of the real and objective impact on the overall situation.

In Figure 10.3 you can see a real example of an ATRED. The item in line number 5, "Supervision of investment projects" involved an average of four items per week. And all in all, this did not seem at first glance to be an excessive amount of time spent, given the importance of the subject matter. However, when it was compared with other activities, such as the one in line 6, "Participation in strategic projects," which totaled 4.5 hours per week, and considering that the person in question already worked an average of 50 to 60 hours per week, it was realized that the same number of hours was being spent on activities of a different nature and relevance. Moreover, in the present case, the situation was aggravated by the realization that he did not have the time to perform other activities considered strategic. The decision taken in the example just described was to reduce the supervision of investment projects from 4 to 2 hours, and only at decision-making times, and to increase direct participation in strategic projects from 4.5 hours to 9. Meticulously following

Figure 10.3 Example of a real ATRED.

the same approach, activities totaling 15 hours per week were reduced in this example, of which about ten hours were transformed into hours available for new activities deemed to have a greater strategic impact, while the remaining five, about 1 hour per day, are dedicated to leisure activities outside the company.

Figuring out what to reduce or what to eliminate is much easier when you try to rebuild the entire jigsaw puzzle of activities and commitments that fill your time. By looking at the entire jigsaw puzzle reconstructed, you will become more aware of what you are sacrificing the most or what you think is important—Gold Activity—but has no place in the intricate garden of your thousand commitments.

It may initially seem difficult to reduce the time spent on certain activities, but the great thing you will discover is that **if you never ask yourself the right question, you will never find the answer**. If you don't ask yourself "how can I halve the time spent on this or that activity?" The answers will never come, neither from you nor from the co-workers you might involve in this analysis.

Among the activities we must sift through are certainly all those that are apparently small and insignificant, but nevertheless repeated many times over time. Reading e-mails, for example, is a recurring, frequent, apparently insignificant activity, but which ultimately takes no less than 1–2 hours out of our days every day, which means about 5–10 hours every week, or 20–40 hours every month, or 240–480 hours every year: 30–60 working days every year. This time does not immediately emerge on a conscious level because in the instant we experience it, we do not have an exact perception of the time spent distractedly reading one or more e-mails. But these times add up, very often multiply, also because we reread and "handle" the same e-mails several times, and in most cases, they create a chain of lost time on other activities that we interrupt because of those very incoming e-mails. Among the solutions that I have most successfully seen adopted are those of "batching," bundling together small, necessary but "time-wasting" activities, and "slotting," devoting isolated, highly concentrated fractions of time to one topic at a time, both of which apply to e-mail management and all those recurring, repetitive, and apparently irrelevant activities: checks, signatures, authorizations, etc. Among these activities I also include the famous breaks so common in companies of the type "do you have five minutes?" Breaks and interruptions that can come in person, via e-mail, via WhatsApp, via phone. Often, they come from the outside, but just as often they are self-induced. Just think of all the times when we feel the irrepressible urge to make a jump on the social network of the day, the inbox, the web. Deny yourself the pleasure of escapism or satisfy the irrepressible need for novelty and telematic pleasure? No, simply take it in controlled, pre-planned doses. You will realize how much time pops up that you didn't think you had.

In the doubt that I will attack you, have a couple of "wild card" questions ready: "what is the worst thing that can happen if I eliminate this activity?" And immediately afterwards, "what can I do to minimize the undesirable effects of what I will eliminate?"

I suggest that you repeat your ATRED periodically, because this brings with it the acquisition of another key habit: to stop and reflect and monitor intentions and actions in a structured manner. There are those who have found it fruitful to stop every month and those who have limited themselves to do it once a year. The choice is yours. Human beings have a thousand ways to tease, sabotage themselves and send back to infinity what they consider to be right and important, overcome by the primordial fear of really changing anything and overwhelmed by short- to short-term logics that take the place of medium- to long-term visions and planning. Once you have established what to reduce, eliminate and delegate, act and monitor what happens after each experiment. Dare. You will be rewarded for your courage!

Testimony

"I realized that I was living my role as if I still had a head of function; I was carrying out everyday activities without giving real priority to 'important' activities and I delegated little. I reflected on my Gold Activities and developed a series of actions that allowed me to reduce waste and time to devote to gold activities.

At the beginning I was working almost 60 hours a week and value activities were only 27%.

There are two main directions:

■ Full delegation of non-strategic testing and reporting activities
■ Slotting of routine tasks (e-mail, activity coordination, scheduling, meetings with suppliers)

In six months I increased Gold activities and reduced working hours by 10%. They are now 46% of my working week. Through the reduction of many wastes, both in terms of the quality of the effort devoted to the different activities, I have gained 6 hours of time, increased my output, and lived my daily life with less anxiety. The positive effect has also spilled over to my team, which now works better and with more results.

What did I learn?

- The hours of time recovered I can spend in strategic activities, this allows me to have a top-down view, not to suffer but to manage events, prevent problems and give myself specific goals.
- "I realize that there are some activities that deserve more attention and time than others. I can read my surroundings better and I adjust my commitment based on importance (I used to devote the same attention to different activities).
- Thanks to a weekly dating routine I am able to pass on information better and receive it.
- Team spirit: I work more with others, I am less jealous of my work
- I have more fun doing it, perhaps because I am less tense and experience everyday life with less anxiety (for example the Operations Committee, industrial test result); I have less anxiety and pass on less anxiety to the guys in production; I sleep more".

—Monica Redenti, R&D Manager NT FOOD

Testimony

With my first round of ATRED, I gained a great deal of peace of mind because I eliminated the feeling of oversaturation and gained time to clarify my value activities on a daily basis. In this change, I enlisted the help of my co-workers and constantly asked for feedback, so that I could refine my ATRED and better understand what my Gold Activities were. Any advice? "Refocus on your ATRED periodically because it is a process by which you can find new obstacles to remove in order to achieve more results and greater well-being.

**—Alessandro Trivillin,
Chief Executive Officer Snaidero Cucine**

Lean Lifestyle Story: Lucchini RS

The Sum Makes the Difference

Striking results are often the result of the sum of many small different actions that, coordinated with each other, change the way of working and thus the structure of one's working week. One should not expect that a single action can change situations settled over years of work.

Mario Guarino, maintenance manager at Lucchini RS, a company operating in the steel and railway industry, gave an example.

With his ATRED he had the goal of becoming more effective at work and making more time for aspects of his private life. Not only that, in fact, it was the norm to come home late every evening, but he continued to work with e-mails that chased him even outside the company and on weekends.

The first step was to focus the gold activities and a careful reflection allowed him to have very clear ideas.

A key first habit he introduced was the weekly reflection slot with two objectives:

■ Doing tactical planning for the coming week: what is open? What needs to be done to meet the deadlines? What tactical changes are needed in the agenda?

■ Reflect on the improvement: how was the week? How did it "work"? What went well? What should I do differently to do better?

With this frequency you can reflect on your ATRED, and the actions taken, significantly increasing the success rate. Mario took the advice of placing this time slot on a Friday afternoon: a time that is typically quiet and good for "entering" the weekend having put all pending chores in order.

> I derived an important personal benefit from it," he says, "because it calmed my mind and prevented thoughts of work issues and to-do lists from emerging over the weekend, which inevitably spoil the best moments of the weekend.

Mario then focused on the main wastes he observed in the initial phase of analysis of the current situation: meetings, slotting, e-mail, delegation. Let's look in detail at the actions he took he introduced standard agendas in the meetings and defined precisely what the focus of the meetings was and for those he governed directly he standardized the duration (45' + 15' buffer, reducing them from the initial 60').

By analyzing the contents of some recurring meetings, he also noticed some duplications and rationalized them. For example, he incorporated into the weekly environment and safety meeting the topics that were previously dealt with in two other meetings in a very dispersed and ineffective manner, succeeding in reducing the total time spent and increasing decision-making effectiveness.

He then noticed a very specific personal waste: meetings were often ineffective and ended up with nothing because he did not come prepared enough. Root cause: he did not plan time to prepare and swallowed by everyday emergencies he could never find the space to do so. He incorporated the solution into his week's slotting: every Monday and Wednesday immediately after lunch he put in two fixed 45-minute slots for the preparation of all the non-repetitive meetings of the following days. Thus, he not only managed to arrive at the meetings focused, but also managed to avoid missing out on important things to bring to the meeting or important information he needed to collect.

He then redesigned the slotting of his entire week by introducing, in addition to the weekly *hansei* and meeting preparation slots, a few slots of sacred time in cadence to work on Gold Activities.

The work on e-mails then led, after a few weeks of work, to a sharp reduction in the time devoted to him, from 12h/week to 6h/week. How? By concentrating reading in dedicated slots in which to work on it (three was his number: morning, after lunch and in the evening before going out) and at the same time learning not to be distracted by e-mails at other times of the day; and then some work on the e-mails received, especially in cc.

Finally, he delegated some non-core activities for his role to a few staff members: in this case, the transition was not immediate but required the structuring of a delegation plan that he started to develop and that will take several months to be finalized.

"The method has served me a lot: the ATRED has now become my working tool," he emphasizes.

> Two things helped me most of all. The desire to learn something new, to see if it could benefit my business even if it meant taking a different path from the usual and from what I thought was best until yesterday, without fear of making mistakes. And then having learnt to give space for reflection, for thinking about what you do and not just doing it.

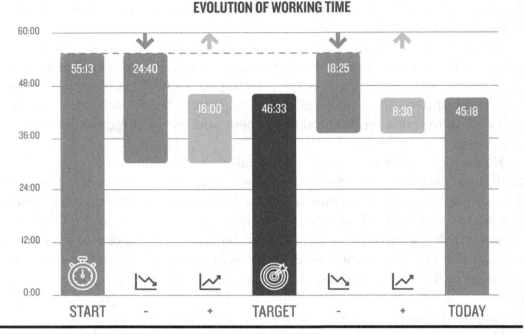

Figure 10.4   ATRED results by Mario Guarino.

Mario's project continues, but the final results already after 3 months of work have been significant: more than 18 hours per week eliminated thanks to the reduction of waste (about 3.5h/day) and 8.5h of the saved hours reconverted to Gold Activity that he did not do before or did not do enough.

The number of hours worked per week was reduced from 55 to 45, which gave him more time for his private life.

Lean Lifestyle Story: Ferretti

Where Are You Running?

Thinking about Gold Activities made me wonder: where are you running? And from here I started a process of structured delegation process of activities that I used to do. For example, I used to spend half a day in production, checking the progress of the boats, if everything was in order, but I ended up doing "instead" of the line manager, many activities. I was always driven by urgency: if there was any problem with a boat, I would go.

Ivan Polidoro, plant manager in Ferretti Group, Italian excellence in shipbuilding, has used the ATRED to challenge himself profoundly, reaping changes in his working methods that were unthinkable at the beginning.

> The very first thing I felt the need to do was to share with my parents. In small doses, but I started telling them everything and became a teacher. This helped me a lot to understand the concepts better and above all, it made me make a commitment to them and made it harder for me not to comply. Today I obviously go into production, but in a structured way: I planned several different gemba walks during the week and the result has not only been good for me, but also for my co-workers who have grown and for some it has meant a lowering of the pressure and stress level due to my approach to work.

Delegation as a resource to achieve more results and greater well-being in the company has been taken to a very high level by Ivan, so much so that in the plant he manages, he has even experimented with what is sympathetically nicknamed the "impossible delegation" in which a high-potential "delegate" is periodically identified to give him the role of a superior for a short time. Ivan tells us:

> For example, in my case, a recently arrived employee, the day before a team-building activity that would take my entire front line out of the company, was heard saying "tomorrow you will be the plant manager." Except for this case, which I remember fondly, we usually take two weeks to prepare them by explaining the essentials, and this has accustomed us all to focus on the important things that really make a difference. Generally, everyone is amazed at how well they do; the ritual reflection is always "but you thought you had these skills?"

After time and constant revisions of his ATRED, Ivan Polidoro admits:

> The Lean Lifestyle was a real enlightenment for me: I had applied lean manufacturing in the factory for years, but I had never thought of applying the same logic to myself and my time.

Among the actions realized by Polidoro and his team:

■ Definition of "individual time" every morning (sacred time).
■ Complete redesign of the line meetings, which went from 3 hours to 1 hour (where everything became "visual," line managers arrive prepared with their critical issues, PMs with boards with weekly planning).
■ Deleting notifications in e-mails.
■ Reducing the flow of e-mails with careful use of "cc" by employees.

"The most obvious sign that I had managed to make a breakthrough in my Lean Lifestyle® project was when I started coming home early," Polidoro concludes. "Before, I used to come home late, even after 10 p.m.: I really had the work mule syndrome. It was challenging, but in the end, I was able to do things better and in much less time. One thing I'm proud of? We were able to close the plant on Saturdays while maintaining productivity. Also, when Covid arrived, we were ready: the team was clear in the preventive phase, and we were able to keep our heads up and thus mitigate the impact. A great success."

Note

1. https://hbr.org/2011/06/the-happiness-dividend

Chapter 11

Become Results Experts in the Smart Working Era

Something Went Wrong

More than a hundred years ago, in January 1914, Henry Ford put in place a real revolution—still little understood today—in the industrial and economic world: he reduced the daily work shift from 9 to 8 hours for the employees of his automobile factory in Piquette (*Detroit*) and simultaneously increased the daily wage from $3 to $5. "We have resolved to pay higher wages in order to create solid foundations on which to build the company," explained the founder of the eponymous car company.

> Five dollars for an eight-hour working day was one of the most effective cost reduction strategies we have ever implemented. So high wages to get low costs. Historians and analysts, studying the discipline, loyalty, and efficiency of the Detroit automaker's workers a few years later, proved that this was indeed the case. So much so that, according to the US economist Gregory Mankiw, Ford "discovered the efficiency wage theory and exploited high wages to increase labor productivity."[1]

Although the Fordist business model generally takes us back to the stereotyped image of the standard eight-hour working day and the concept of the extreme fragmentation of production, it is paradoxical to note that Henri Ford's most radical intuition was precisely to decouple the logic of

DOI: 10.4324/9781003474852-11

productivity and wages from the hours worked. An intuition supported by technological, organizational, and social considerations. In fact, the rapid technological and organizational advances allowed the reduction of hours worked at the same time favoring an increase of productivity: the concept of the moving assembly line, the first forms of automation, the integration of component supplies, made it possible to realize Ford's dream of making the automobile accessible to as many people as possible—to the average American—and even to make it much more convenient and functional than its predecessor, the horse-drawn carriage, and then the other cars on the market, which were typically very expensive. The increase in wages went in the twofold direction of increasing the purchasing power of working people and at the same time attracting more and more workers from agriculture and other competing businesses, motivated by the higher wages available and fewer hours worked.

Last but not least, the goal achieved was to reduce absenteeism, which fell from 10% to 2.5% in just two years, and turnover, the ratio of dismissals and resignations to total employment, which fell from 31% per month to 1.3%.

Ford Motor Company exploded, becoming the number one car manufacturer in the world in just a few years, emerging from thousands of competing companies, crumbling down the sales prices of cars and setting new paradigms for the entire manufacturing industry.

It felt like the beginning of a new era. In fact, struggles and disputes over the reduction of weekly working hours had been the cause of trade union conflicts over the years, as well as the chimera of a compromise that had never been reached between technological progress and the reduction of working hours.

Already in the summer of 1930, in the midst of the Great Depression, the economist John Maynard Keynes wrote: "For we have been trained too long to strive and not to enjoy." The British economist predicted that, in the absence of dramatic conflicts, budgets and development would no longer be an issue by 2030 and that Western living standards would at least quadruple. Within a hundred years, the working week would also be reduced to a maximum of 15 hours, thus enabling everyone to devote the rest of their free time to pleasure and health.[2]

It is now less than ten years away and it seems that this prediction is far from coming true, except for global productivity. Indeed, in many countries, the overall production of goods and services has grown dramatically, but leisure time has not increased at all for working people. On the contrary, **the belief seems to prevail that having a little extra free time is almost**

comparable to being unemployed, in a general view that puts work before life and not vice versa.

Something went wrong.

Other economists shared variants of Keynes' prediction, arguing that as advanced economies became more productive, people could have happily chosen to work fewer hours. Unfortunately, this also did not happen.

Despite the impetuous technological development of the last hundred years, we are firmly anchored to the concept of work of the last century as far as the dynamics of the working world are concerned. We verge on the thresholds of paradox in the managerial and entrepreneurial world, where long working hours now even blend in with everything else thanks to—and because of—digital hyper-connection and the flow of communication in all moments of our life. It is, unfortunately, common to see women and men entering factories and offices early in the morning and leaving late in the evening, perhaps with a sense of guilt every time they have worked a single hour less than a standard carved in our minds, even before company contracts. Outside the company, we rarely actually disconnect. And the remote working, which we have been getting used to since the 2020 pandemic emergency, is making the space-time distinction between work and everything else in our lives even more complicated.

The paradox in all this is that we never feel guilty about not being as productive as we could be. We never wonder if we could achieve more – with more quality—in much less time.

Maybe we feel reassured, and often a little stunned, by the fact that we have worked 12 hours straight, but we hardly ever feel guilty for not having produced the right result in the shortest possible time. Yet the technology at our disposal today can really help us in this regard, provided we know how to govern it with clarity. Goods for sale at a high price these days.

Our Mind Behaves Like a Gas

Work expands so as to fill the time available for its completion; the more time, the more important and demanding the work seems.

—**Cyril Northcote Parkinson**

In 1958, the British naval historian Cyril Northcote Parkinson published a small book, *Parkinson's Law,*[3] which unexpectedly became a worldwide best seller that has been translated and is still sold worldwide.

Imagine you have a certain amount of gas at your disposal, and you want to put it into a container. The gas will immediately fill all the space at its disposal. If you want to take the gas out of that container and put it into a larger one, the gas will easily adapt and thank you for the convenience of the extra space you are giving it. You can repeat the experiment at will using larger and larger containers, but you will always get the same result: the gas will expand until the entire space provided is evenly filled.

The mind behaves exactly like gas, it magically occupies all the space at its disposal, in work and in life.

All of us have studied and prepared for some exams, but let us remember: when did we make the greatest effort? Immediately before the fateful date, a few days, and a few hours before. In these terms, some scholars have named this which I am describing to you as the student syndrome, or the last-minute syndrome, all the situations in which we end up reducing ourselves to the last available minute in order to complete an assigned task on the now tight deadline. Our mind behaves like gas much more often than you can imagine.

Cyril Northcote Parkinson was convinced that those who set out to complete a task in just one hour, magically succeed. If you had two hours instead of just one for the same task, it would take exactly two hours! And if you had three available? It would take three.

This concept is of fundamental importance in all situations in our corporate life and in the new situation of remote working, where we always tend to occupy all the time available to us. Planned meetings of two hours always end up lasting two hours, except for delays and handfuls of minutes stolen from the next meeting. Much more rarely do we see meetings that last only half the planned time. Activities that always end up taking up the whole day until dinner time or any other convention that requires us to stop and resume the next day. We could use Parkinson's law and the student syndrome in reverse, so as to enjoy the possible enormous benefits, as Eliyahu M. Goldratt suggested in his enlightening book *Critical Chain*: halve the time available to reduce work to the essentials and have no time for distractions, multitasking, loss of focus.[4] We would discover the power of concentration, of the deadline adrenaline that plays in our favor to get the task completed by the tight deadline. If it is true that our mind is like a gas, we could experiment with putting it into smaller and smaller containers until we reach the maximum compression state of the gas, the state in which it turns into liquid. The final product of the closer deadline will almost inevitably be of equal or higher quality due to concentration, also because the **perceived importance and complexity**

of a task increases in relation to the time allotted for its execution. We would discover the magic of the impending deadline. Limiting tasks to the essentials to shorten work time and shortening work time to limit tasks to the essentials are the mantra we need to equip ourselves with to protect against our innate tendency for easy expansion.

What If Working Less Actually Made You More Productive?

During the 2019 World Economic Forum in Davos, Rutger Bregman, a Dutch historian, and economist, created some embarrassment with his speech on industrial countermeasures to avoid crises and social repercussions. Among his proposals was to reduce working hours to 15 hours per week to really increase the productivity of the entire system, just as Henry Ford did a century ago by reducing working hours from 60 to 40. This idea had already been proposed in one of his interesting books, Utopia for Realists, and taken seriously by several companies.[5] The New Zealand estate planning company, Perpetual Guardian, confirmed in 2018 that it would make permanent the experimental measures it had adopted for eight weeks on its 240 employees, increasing the number of days worked during the week from 5 to 4, for the same salary and conditions of employment. The trial had confirmed a 20% increase in productivity, significantly reduced stress levels, increased job satisfaction and improved work–life balance.[6]

And in its permanent extension, the four-day work week was opened to forms of flexibility of the employee's choice: work four days a week or fewer hours per day for five days a week or start and finish earlier to avoid traffic congestion and spend more time with the family. The founder and CEO, Andrew Barnes, after studying the many productivity analyses done on his company, stated in 2018 that this would be a form of modern crusade against traditional thinking and standard paradigms regarding work rules: "Working hours are not all the same. Long working hours do not equate to greater productivity!" Barnes, in fact, even created a non-profit group to promote the four-day week, and New Zealand's Prime Minister, Jacinda Ardern, in May 2020 publicly declared the link between these corporate measures and the significant growth in the domestic tourism turnover. Andrew Barnes' experiment did not remain isolated in New Zealand, so as not to leave indifferent even Unilever. The multinational company decided to conduct a 12-month trial in its New Zealand headquarters, where employees will be able to work only four days out of five on the same pay. The trial will analyze productivity data to decide whether to extend this model to all its 155,000 employees worldwide. Nick Bangs, managing director of Unilever New Zealand, said in an interview with the Financial Times on 30 November 2020[7] to be aware of the difficulties of applying this model to the production departments, but at the same time to be convinced of the possible benefits that can be achieved in all other departments of the same company.

Although many critical voices have feared that such a revolution could send many companies at a loss, a Big-Company like Microsoft experienced the same route in August 2019 in Japan, known for having the longest working hours in the world. The initiative was created to promote a healthier work–life balance and was called the "Work Life Choice Challenge" and involved the 2,300 employees of the Tokyo office. Again, the proposed model was four days a week. If the recorded reductions in corporate fixed costs due to the absence of operational activities in the offices are understandable – for example -23% in electricity costs and -50% in paper consumption—the productivity figures reported by Microsoft itself for the test period are less so. Productivity, measured in terms of sales per employee, increased by almost 40% over the same month of the previous year.

The pandemic has definitely shaken up traditional thinking on the dynamics of work and perhaps we have taken a path without return. "Let's imagine a hybrid working future, where people could spend a couple of days in the office and two or three days at home or work remotely," said Alan Jope, CEO of Unilever. "This situation and this new way of working have paradoxically unlocked incredible productivity and flexibility throughout the Unilever team."

All over the world, the number of companies applying partially or fully applying the WFA model—Work from Anywhere[8] (work anywhere from a distance) is increasing- bringing to light both new critical issues to be addressed and new opportunities to be seized, for companies and for individuals. The Covid-19 has only given a considerable acceleration to a movement that was already in progress.

And what is the situation in Italy? What are companies doing to improve both results and well-being? What have been the effects of Covid-19? We will see this in the next paragraphs.

Mere Work Does Not Ensure Results and Well-Being

In these months, due to the Covid-19 pandemic, we have all been forced to experience remote work or distancing in the offices. The subject of space-time boundaries is even more pressing, and many speak of the risk of encroachment of the working sphere into the personal one. In my opinion, the greatest risk is not this: the biggest problem is not being able, even more, to distinguish the results we produce through our work, from the work itself and the many activities we do every day. At home everything is made even more complicated by the presence of other sources of domestic distraction and the absence of confrontation with our colleagues.

A few years ago, Cali Bessler and Jody Thompson published a book that is now relevant again and which provoked controversial reactions at the time.[9] The authors argued very strongly that the traditional concept of work—at least 40 hours per week, Monday to Friday, 9 a.m. to 6 p.m.—was outdated and that it was unbearable to see *"people who are inadequate for their tasks being promoted just because they arrive earlier and stay longer than everyone else at their station."* The solution was the construction of a work model focused only on results and not on activities, *ROWE—Results Only Work Environment*—cancelling the concept of work as a physical place and transforming it into a metaphorical place where transparent objectives are agreed upon and exchanged between employers, managers, and employees. Hence evaluation only on tangible and measurable outputs, no longer on activities performed. A revolution shouted in advance of today and perhaps all to be contextualized, but certainly worthy of attention and deep reflection.

One thing is certain. We live in the so-called Information Age, the technology that brings people together, but in essence the nature of the workplace, the hours, and the obligatory presence behind a desk have not changed compared to the Industrial Era, when the assembly line demanded the physical presence of the worker and all the people responsible for issuing work orders and related controls.

Remote Work Trials before Covid-19

Even before the Covid-19 pandemic, many companies were progressively and increasingly implementing extensive forms of *Smart Working. For* example, Siemens was one of the first companies in Italy to implement a *Smart Working* project, starting in 2011. At the beginning of 2019, 1,700 employees of the group were working in agile mode, without clocking in and without a fixed desk. Other companies, such as Fastweb, Microsoft, Vodafone, Accenture, Unicredit, Intesa Sanpaolo, Bnl, Ubi Banca, and Tetra

Pack, have experimented with different forms of remote and flexible work, collecting interesting results in terms of increasing productivity, reduction of fixed costs and physical space management, and reduced absenteeism rates. Observations and data meticulously collected by the Smart Working Observatory of Milan Polytechnic.[10]

Barilla has also been one of the most active companies in this field in Italy. Since 2013, it has launched a project with the aim of giving employees the opportunity to work flexibly, anywhere and at any time, thanks to new digital communication tools and new methodologies. Around 1,200 employees have taken advantage of the opportunity. Smart Working in Barilla was open to the whole population—from a contractual point of view: employees can work in locations other than the office for 4 days a month, agreeing with their manager. Overall, the outcome of the initiative has been extremely positive: a greater work–life balance has made employees more satisfied, thanks to greater autonomy and responsibility over when, where and how to work. And the cases have not remained isolated: according to data from the Milan Polytechnic Observatory, in fact, in Italy the new approach to work organization has been introduced by almost one in two large companies in a structured or informal way.

Smart Working? A Violent Change with Uncertain Outcomes

Ironically, the Covid-19 pandemic and the various lockdowns have disrupted the plans of both those companies that were moving in the smart direction and those companies that had not yet begun to think about it. Our usual ways of working have suddenly changed as have the smart conversion times of all companies, violently imposing new forms of remote interaction and forcing us to accept tools already known to many, but never used so massively: data and information sharing platforms, video meeting solutions and video distance learning. During the 2020 lockdown, 94% of PA, 97% of large companies and 58% of SMEs extended the possibility of working remotely to their employees.

The change was sudden. Physiologically we will see a descent and normalization of the phenomenon at the end of the emergency, but it is widely believed that it will leave irreversible marks on the world of work. According to the estimates of the Milan Polytechnic, we went from 570,000 Italians in agile work before the Covid-19 pandemic to 8 million remote workers during the lockdown. All in the space of just a few weeks. But there are other interesting data, such as those which emerged from a survey by CGIL-Fondazione Di Vittorio: 94% of those interviewed agreed that smart working saves time lost every day on the home-work-home route, allows more flexibility, gives the chance to work effectively by objectives and, last but not least, enables a better balance between work time and free time. At the same time, **71% of those surveyed stated that working from home suffers from a lack of opportunities for discussion and exchange with colleagues** and, in addition, increased family burdens. A willingness to continue, perhaps with a mixed formula, emerges in 60% of the answers. We will see. To date, the criticalities of remote working, superficially referred to by many as smart working, are still many. It is not enough to stay in one's pajamas and connect to the rest of the world with a PC and a good Internet connection to determine its effectiveness. Surely there is that many people suffer numerous problems resulting from the immaturity of our infrastructure and technological system: 11.5 million users in Italy are not yet served by a broadband connection. And even those who are theoretically served, suffer performance variations and disturbances of different kinds during connections. Without bothering with the various researches and numerous surveys that have proliferated during the pandemic period, I think we have all experienced first-hand the difficulties in communication due to connection problems, audio-video quality, delays in setting up and starting up video meetings, not to mention the difficulties in staying focused, without being distracted by other temptations, even more facilitated, during the meeting itself: e-mail, social networks, websites. According to data from the American company OWL Labs,[11] during the year 2020, despite the many benefits of working remotely, the problems just listed above have led to an extra workload of around 26 hours per month, that is almost one extra working day during the week, and a 50% increase in time spent in video meetings. Not to mention the ability to concentrate during video conferences, which already tended to be difficult even in presence, dropped by around 30% according to Gartner data.[12] In addition, many companies underestimated the cyber risks, so much so that they were helpless in the face of a surge of malicious attacks hunting for data and information during the lockdown weeks.

Another critical element is that many companies, public and private, have sent people home, but have not even taken the time to design new rules and processes necessary to make remote working effective. Many people have paid for their inexperience at work by goals: if you have never done it, it is impossible to improvise alone at home, perhaps by staying for a long time connected in front of a screen in endless meetings.

Self-organization skills, as we have long examined in the first chapters of this book, become essential to minimize time loss, increase one's productivity and regain energy, but in remote work situations, these skills play an even more important role. In fact, in the so-called remote work and in the particular context of pandemics over such a long period of time, it is very easy to get into cognitive and emotional states that make it difficult for us to make a lucid choice of what is right for us and for our productivity.

Our way of working was already critical in the beginning, we saw in detail in Chapter 4. But the pandemic situation has worsened the situation and demands even greater control and lucidity. The flood of information that inundated us and that we continue to receive has unfortunately had several unexpected effects on us. It has shifted our focus to elements external to us, on most of which we can do little or nothing, except for personal protection, distance, and isolation. And we have been projected, even more, into a working condition reactive brain. It is as if our brain, already hyper-stimulated in its remotest parts by the media bombardment, were to enter further vicious circles and continuously search for more and more up-to-date information, perhaps in search of reassuring news. We unconsciously lose conscious control of time and find ourselves in a subtle state of anxiety that affects our working days from home. The various meetings we attend often find us less concentrated and in a state of poor intellectual performance, which in most cases makes the smooth running of the meetings even more difficult. We risk further accumulating delays, stress, losses of individual and team productivity, and a deterioration in overall effectiveness.

The last critical element to be addressed in remote working concerns the ability to effectively supervise employees not at their desks, an issue that is already difficult to face in "normal" situations.

How do you guide others when I don't see them most of the time?

How do you give feedback when you don't have the opportunity to look into a person's eyes and observe all their body language, pick up on non-verbal and para-verbal messages during a cold communication behind a PC screen?

What skills do I have to stimulate and develop in my employees in smart working?

How do I align my team, how do I share goals, critical issues, and countermeasures?

What kind of leader do I have to become in this situation?

One thing is certain. Until a necessary two-step cultural and system is made for a truly smart work model, working from home does not automatically ensure more individual and company results, less activity, fewer hours worked and above all does not ensure more well-being. On the contrary, in many cases, it will save us a few hours of homework travel, but worsens the situation overall. Unless we take new paths and acquire new skills.

Testimony

I have been working from home for 13 years now for PPD, a multinational company with over 24,000 employees, specializing in contract research that provides comprehensive, integrated drug development and lifecycle management services.

I remotely drive a team of 12 to 15 function leaders located in every corner of the planet. Working in this mode and having at most once a year the opportunity to meet my co-workers in person has led me to develop a new leadership style, and in this I have certainly been helped by the Lean Lifestyle Leadership model.

Remotely, the distance between reality and perception is in fact dangerously thin and it is more difficult to build relationships based on everyone's authenticity. For this reason, as soon as I start working with a new team, I dedicate time, with specific one-to-one meetings, to laying the foundations for a relationship that does not move from mutual perceptions. I ask a lot of questions investigating what one likes and dislikes doing, what is and what is not important in one's personal and professional life, and what one's strengths and weaknesses are. It is not an easy task because it has to be calibrated on a case-by-case basis, as I am faced with all kinds of characters, personal stories and cultures: sometimes it helps that I am the first to highlight my weaknesses or the elements of me that I would like to develop.

It is about finding the "key to access" for everyone and then the rest comes as a result, maintaining a leadership style that is never judgmental, constantly listening, and aimed at focusing people toward objectives and problem-solving. People in my teams must not be afraid of making a mistake, because when this happens, I must ensure that they have the freedom to focus on solutions. I am the one who exposes myself to customers when there are difficulties and I let them live their work in peace.

The Lean Lifestyle Leader approach has contributed to my growth and the results of the teams I manage. If I had to give advice, I would definitely say to never judge the person in front of you and to never stop asking yourself questions to improve your leadership by taking on board the point of view of your co-workers.

—Veronica Bini, *Director Oncology Project Manager PPD*

From Smart Work to SMART Goals

Remotely, inside our homes or wherever we want, or in presence, within the physical boundaries of our company, the number of hours worked will lose more and more importance in the criteria of assessment and assignment of professional tasks and duties, while the objectives that are assigned to us and that we in turn assign to our co-workers and project teams will become increasingly important. For this reason, I consider it important to conclude this section of the book with an in-depth look at the criteria for the correct setting of objectives.

For this purpose, I will use the acronym SMART (Specific—Measurable—Achievable—Relevant—Time-bound), which first appeared in the *Management Review* in an article by George Doran in November 1981.[13] Used throughout the world of education and management, this acronym has the unquestionable advantage of being easily understandable, recallable, and above all recognized when put into practice, so much so that it has been commonly associated with the "management by objectives" criteria developed by Peter Drucker,[14] a world-renowned author for his works on business management theories, as well as a professor, economist, and esteemed consultant for hundreds of companies of all sizes. So, back to us: a goal must be SMART when it is specific, measurable, achievable, realistic, and timed. Let's see in detail.

Why Should a Goal Be Specific?

A vague goal leaves room for ambiguity, whereas a specific goal clarifies the area of action well and provides simple, easily understandable indications of what is to be achieved. If, for example, I wanted to set the goal correctly regarding the writing of this book, I could have stated that I wanted to write a book. Does that sound specific? Not enough. A book can be of different literary genres; it can be written as an essay or as a manual, it can be a novel, it can be popular or for insiders, and it can cover different fields in both business and academic. I decided to write a book that collects and disseminates the best business and managerial practices that are useful to realize a new business model, a Lean Lifestyle Company, capable of achieving great performance and great well-being at the same time. A book that rewrites the traditional way of working. A book intended for entrepreneurs, managers, and professionals interested in transforming themselves and their business organizations to work more streamlined, more agile, and achieve more results with less stress. Much more specific as a goal than just wanting to "write a book." Let's move into the company. A desired goal, but definitely badly placed, could be the generic "we must improve our level of customer service." A more specific corporate goal could become: "we want to improve our response time to customer enquiries" or "we want to improve the turnaround time of customer orders received," focusing instead on the scope of the executive delivery of the product or service we offer to the market. Two specific objectives that will give rise to two separate projects.

Why Must a Goal Be Measurable?

A measurable objective can be assessed both at the completed task or project and during the ongoing activities. Taking my book as an example, the objective defined above becomes measurable when I add the number of pages I want to produce, the number of companies I want to include as case studies and testimonials, and the number of copies I want to sell. In the case of this book, I have set several pages between 280 and 320, a minimum number of case studies and testimonials of 20, a quantity of copies sold of at least 10,000 in the first two years. All of which can be easily measured by anyone. Returning to the example of the customer service level, we will add specificity to the objective when we define the two objectives above in these terms: "we want to improve our response time to customer enquiries, reducing it from the current 4 weeks to 2" or "we want to improve the turnaround time for customer orders received, reducing the average value from 21 to 10 days."

Why Must a Goal Be Attainable (Achievable)?

The letter A in the acronym SMART in its initial version stood for *assignable*, which is clearly assignable to a person who would take full responsibility for the objective. And all in all, I don't mind at all. However, taking it for granted that goals in business are born to be assigned, I agree with the common and now widespread use of the word achievable—attainable—for the third letter of SMART. In the case of my book, I could have written 1 million copies sold as a target, but evidently the probability of reaching this goal would have been very low, so low that I would have had to mobilize resources and energy not only at the Italian level, but internationally, to increase the concrete chances of succeeding. After all, anything is possible, but I preferred to set a challenging and at the same time attainable goal in the time available to have activities compatible with my other commitments as a father, husband, entrepreneur, consultant, and CEO of my company, without sacrificing my well-being. In the case of the example of the two business targets, on the other hand, someone—as I sometimes hear—might have *wanted* to fulfill all customer requests in a few hours or a few days, or almost immediately or in any case in such a timeframe that it would not be perceived as realistically achievable. It is a different case when the target is set based on observing the data, perhaps discovering that a part of the requests is dealt with in less time than the average, and therefore focusing attention on how to constantly replicate a result already achieved other times in the past.

Why Must a Goal Be Realistic?

A goal can be achieved conceptually but proves to be unrealistic if not translated into concrete terms in the real life of people and organizations. Very often, ambitions and desires are mistaken for the real possibility of doing what is necessary to achieve the goal I have set myself. I may have critical issues in terms of personal time, or the resources allocated to my project, but it is crucial to be very objective when setting a goal. For example, my goal of writing the book, described so well so far in terms of specificity, measurability, and attainability, was a dream unhooked from reality until I decided to commit and put in the agenda at least 15 hours each week to devote completely and solely to writing. Like the other examples of business objectives on service level improvement, they do not become realistic until I am certain that I can commit portions of my own and other people's time to devote to project design, analysis, countermeasure study and implementation. So, whenever you want to make your goals realistic, check the actual availability and allocation of time, resources, money, as well as any constraints to overcome and permissions to request.

Why Must a Goal Be Timed?

A goal can be considered timed when it is unambiguously linked to the relevant timeframe and the achievement of intermediate outputs along the way. The goal of writing the book became definitively concrete when I set a time limit for both the final realization and the intermediate deadlines: I started at the end of January 2019, estimated 24 months "gross" for the final delivery to the publisher, scheduled for the late January 2021. Taking away 2 months for each year due to holidays, holidays and planned non-writing periods, on average I would have to write about 15 pages per month, excluding the months of August and December, to achieve the result of about 300 pages by the end of January 2021. Going into further detail, since I do not write every day, but about 2 days a week, the expected output to maintain the schedule was set at about 2 pages a day and 4 pages a week. The manuscript will be delivered to the publisher, Hoepli, at the end of February for all the editing, layout, proofreading and graphic processing, to finally go to the bookstore in June 2021. If you go back to the example of the two service level improvement projects, the timing will cover the various project phases with a clear indication of the expected output. This can be an example: list of the main critical issues of the current processes by the end of the initial analysis phase, estimated at a total of one month from the start of the project; list of countermeasures and improvement actions, validated by the team and the plant manager, within 15 days from the start of the future process definition phase; completion of the first wave of implementations within the first month of the execution phase and second wave implementations within the second month of the execution phase.

From SMART Objectives to SMART Results

There is often confusion in the company about the difference between objectives and results. It is not only a question of terminology or semantics, but of substantial differences.

Goals represent the schematic and synthetic quantification of something we want to achieve in the future, which can correspond to our own desire or to an "imposition" of others.

Results represent what we have achieved or want to achieve because of our actions, the actual change that has taken place or is desired.

Knowing how to distinguish between objectives and results will enable you to constantly compare what you wanted to achieve with what you have achieved, and the measurement of this gap will be the engine of continuous improvement that is triggered when analyzing the causes of the lack of results and devising new action plans with new, ever smarter objectives. The result achieved may not be what we were aiming for or what we wanted, and in these cases, we speak of an "unwanted result." How far have we deviated from the set goals? Is the difference between objectives and achieved results significant? Can it be corrected, or should we now take it for granted? Was the deviation due to an error or inability, or was the target too ambitious and does this need to be changed? These are some of the reflections that will flow from the *hansei* process, which we will examine in detail in the next chapter, applied to the analysis of our results. We may find that we have poorly formulated our objectives or that we have not even defined them. We may also find that we have not thought enough about the meaning and purpose of the outcome related to the objective. We may find that we have underestimated some insignificant consequences related to the expected results. To achieve the "desired results," that is what we want to achieve, it is important to formulate our expectations well, in addition to the SMART goal-setting criteria just discussed.

Some Questions to Verify That You Have Formulated Your Own Expectations Well

Is Your Goal Expressed in Positive Terms?

To become an expert on results, we have to express expectations and goals in affirmative terms and pay attention to anything that might be expressed in terms of negating something. Only in this way can we focus on concrete elements that we can build through the different phases of a project, described by smart action plans, programs and goals.

Is Your Objective So Clear That It Needs No Further Explanation?

Anyone must be able to understand it without giving any possibility of interpretation. Therefore, it is not enough to be specific: but it must be written in a simple and clear form so as not to create any doubt in the reader, both internally and externally. This will help you and others, because it will force you to be as simple as possible in your description and combine pragmatism with effectiveness of communication.

What Is the Meaning of Achieving the Goal?

What does it really mean for you to achieve that result? And for your group? For your company? This criterion is powerful because it links the rational sphere to the deepest sphere of people, also touching emotional and "spiritual" aspects. When we reflect on the purpose of our actions and on the ultimate goals, we give meaning and significance to our actions. If we understand why we must do something—the so-called *reason for doing it*—we will focus on both the positive consequences, we want to benefit from thanks to our actions and the negative consequences we want to avoid. In this way we will feed the energy and align the focus of each member of our team.

Is the Achievement of the Goal under Your Control?

Before we dive into the action and to correctly estimate our chances of succeeding, we need to fully understand what is under our direct control and what is not. When something is completely under our control, we have all the levers and resources to act. When it is not, however, the only weapon we can use is our ability to influence, our leadership, and our ability to "exploit" the resources of others. This will lead us to precisely identify responsibilities and resources of other people to whom relevant parts of the project are to be delegated and to establish with them moments of confrontation and monitoring to carry out periodic *hansei* related to their activities.

Do You Have All the Resources Necessary to Achieve the Goal?

With the right resources, you will dramatically increase your chances of getting the desired results. Internal resources: your own and your team's skills, the right people and above all their time. External resources: equipment, suppliers, economic resources and other external sources of knowledge and support. Knowing how to accurately identify the necessary resources and above all knowing how to "exploit" them well will be the right lever to move toward the desired results.

Have You Examined and Addressed the Areas of Influence and Addiction in Advance?

Everything we do has many possible influences, direct or indirect, on other areas of our lives, both on a personal and corporate level. We live in constant interconnection with everything around us and it is good to analyze in advance, before diving into the action, the possible negative and positive influences, induced changes and other side and undesirable results associated with the main desired outcome. In the example of my book project, described above, I faced both positive and negative influence zones. Among the positive consequences: increased notoriety, dissemination of the methods and projects developed and applied, *positive personal emotional return* for doing something I deeply enjoy and have always dreamed of doing. Among the negative consequences I had to put on the scales: reduction of time available for myself, my family, my co-workers and my company, possible increase in stress due to time pressure and the management of tight deadlines to respect. To avoid undesirable side effects due to these negative consequences, I chose—in advance and consciously—to dilute the project of writing the book, to deeply reorganize my company to create more and more autonomous managerial figures over time, and to further simplify my personal and professional life. These choices have led me to delegate much more and to give up or downsize many things that I consider less important, such as television, social life, control and micro-management in the company, secondary projects or those considered non-strategic, in order to focus on taking care of myself, my family, the high-level management of my company, the training of my collaborators and the continuous construction of systems, processes and organization rather than the more executive aspects.

Dulcis in Fundo

In conclusion of this chapter, the final recommendation to become an expert in results and well-being at the same time. Pay attention to your senses. Use them all. Become fully aware of your emotions, of how you really feel. How? By giving yourself a few minutes every day to do this and to focus on yourself by turning down the external background noise and eternal connection. Don't close your eyes to the weak signals that your body and your emotions give you. Don't go on with false optimism. If your senses tell you that things are not going according to plan, use this information to adapt to what you are doing and get back on track. Be flexible in your behavior. There should be no ostrich-like behavior, stubbornness like a horse with blinkers or dogmatism. If what you are doing is not working, do something different. **Be rigid with your goals, flexible with your means to achieve them. You must become smart to work smart.**

Notes

1. Martha Banta, *Taylored Lives: Narrative Production in the Age of Taylor, Veblen, and Ford.* University of Chicago Press, 1993.
   Bernard Doray, *From Taylorism to Fordism: A Rational Madness.* Free Assn Books, 1988.
   Peter J. Ling, *America and the Automobile: Technology, Reform, and Social Change* chapter on "Fordism and the Architecture of Production".
2. John Maynard Keynes, *Essays in Persuasion*, New York: W.W. Norton & Co., 1963, pp. 358–373.
3. Cyril Northcote Parkinson, *Parkinson's Law.* Buccaneer Books, 1993.
4. E. Goldratt, *Critical Chain.* North River Press, 1997.
5. R. Bregman, *Utopia for Realists. How We Can Build the Ideal World.* Little, Brown and Company, 2017.
6. https://4dayweek.com/
7. www.ft.com/content/fc3e1d5a-5abc-4de4-a1b5-b40c5c09620d
8. https://hbr.org/2020/11/our-work-from-anywhere-future
9. Cali Bessler and Jody Thompson, *Why Work Sucks and How to Fix It.* Portfolio Trade, 2008.
10. www.osservatori.net/it/ricerche/osservatori-attivi/smart-working
11. https://resources.owllabs.com/state-of-remote-work/2020
12. www.gartner.com/en/human-resources/trends/remote-work-revolution
13. Specific, measurable, achievable, realistic, timed. These are the definitions, largely corresponding to the original ones of George Doran, which I chose to adopt. Other interpretations have been used over time. For example, Realistic R has been changed to Relevant or Reasonable by other users. If you decide to use my guidelines, the path will be completed with the definition of Results that I will explain in the following pages.
    G. T. Doran, There's a S.M.A.R.T. Way to Write Management's Goals and Objectives. *Management Review*, 70, 1981, pp. 35–36.
14. P. Drucker, *The Practice of Management.* Harper, 1954.

Chapter 12

From Smart Working to Smart Company

What to Do to Make Work Really Smart?

There was a generalized misunderstanding during the months of the pandemic: many mistakenly called smart working the mere work at a distance, forgetting all the other profound implications necessary for the realization of true smart working, that can even be achieved regardless of whether you work at a distance or not.

The company that decides to exploit the full potential of Smart Working concretely and steadily, even when we have archived the Covid-19 health emergency, should undertake a specific development path, and not leave to improvisation such an impactful change on the way of working and doing business, and thus on individual and social performance and well-being.

These are, in summary, the steps you need to take to achieve a smart turnaround in your company:

1. **Consolidation of the hardware and software infrastructure system,** including the design and implementation of cyber-security plans.
2. **Definition of new** organizational **rules**, process reviews, operational principles, and key habits to be adopted in Smart Working.
3. **Definition of new performance management criteria**: work will no longer be evaluated based on hours worked but based on results achieved and consistent with operational principles and key habits.
4. **Design and delivery of a specific training plan**, based on the degree of maturity that emerged: mindset and rules of work by

objectives, new organization and new specific processes, principles of self-organization and personal excellence, key habits in remote working, criteria for supervising and developing employees in Smart Working contexts.

5. **Design and implementation of a specific communication system** to convey the right message from the very beginning and throughout the company's evolution toward its definitive Smart Working set-up.
6. **Design and implementation of a Knowledge Management system** with high usability, completely digital and that promotes the sharing, work team collaboration and the capitalization of best practices, information, and corporate know-how.
7. **Implementation plan** with related monitoring and review steps.
8. **Initiation and run** of Smart Working, monitoring, and continuous improvement cycles.

Leaving aside more or less common-sense advice, about which so much has been written and read over the past year, I will dwell in the next few pages on aspects that are less debated than others and which in my opinion need to be addressed very carefully by companies when planning a transition to Smart Working.

Define Business Rules for Smart Working

To establish effective countermeasures to the criticalities and the framework that emerged from the initial evaluation, it is necessary for the company to formalize and disseminate a series of rules and operating principles specific to Smart Working. I will list ten rules, which represent the synthesis of the best experiences adopted during this period.

1. **Avoid meeting overload.** Eliminate, merge, reduce duration, reduce participants, establish clear inputs in preparation and clear outputs expected from each meeting. Choose to confirm on the agenda only those meetings that are strictly necessary, not being fooled by the ease with which they can be organized remotely.
2. **Establish boundaries and set a weekday with no meetings, neither physical nor virtual.** Openly state time limits for summoning meetings. Establish "free zones" without meetings, for example during lunch breaks and late afternoons, to the point of "blacking out" entire

days. These prescriptions offer the enormous advantage of forcing companies and people to implement the first rule well in order to concretely reduce meeting overload. I repeat work should not be confused with meetings. The latter are only for sharing, aligning, and deciding what to do outside of meetings, in real work.

3. **Apply the rules of slotting and establish regular cadences for thematic areas.** Using the techniques seen in Chapters 5 and 7, avoid filling people's agendas with meetings that pop up like mushrooms, interrupting other activities and preventing "one-task" concentration. Try to encourage the habit of putting priority and important work activities on the agenda in the early morning, to be carried out independently or in very small groups, maximum three people, and try to place all meetings in the afternoon so as to have two conceptually separate bands of the day: in the morning, individual work with high added value—Gold Activity—and in the afternoon, to relieve stress and increase productivity, service activities—video meetings—to ensure rhythm and pace.

4. **Appoint a leader for each meeting**. The person who convened the meeting must also lead it: make introductions, if necessary, grant the floor, bring everyone back to the agenda if the conversation starts to digress. Keep to the timetable to avoid delays. Draw up an action plan before the end of the meeting and share it before the final greetings.

5. **Establish and disseminate a shared method for the management of video conferences**. If the rules are important in conducting face-to-face meetings, they are even more so in remote meetings. It is important for each company to establish and disseminate some simple but "mandatory" rules aimed at maximum meeting effectiveness, for example: make agenda and objectives mandatory, reveal the information required before the meeting, give behavioral rules during the meeting (no distraction and doing something else), set up an action plan and follow-up. This will avoid snags, distractions, inefficiencies and wasted meeting time.

6. **Establish moments of measurement and detection.** The company must be constantly listening to what is happening in the working lives of employees, to detect in real time the level of stress, participation, time spent and other issues that may arise. This can be done in two complementary ways: on the one hand through active analytical tools (short surveys) and passive analytical tools (derived from communication and collaboration tools) and on the other with the classic one-to-one interview between manager and employee, which is valuable and should be made even more frequent, in all situations of physical and social distancing.

7. **Help employees to establish a fixed place to work at home**. Whether it is a small or large space, a dedicated room, or a sheltered corner, it is essential to "carve out" a physical setting that projects the individual into the high concentration working mode. The company should facilitate this change also economically, though, for example, the purchase of a desk, an ergonomic chair, or headphones: it is an investment that will have a double return for the company because it will improve the effectiveness of those who are working and consolidate the sense of belonging.

8. **Help employees to define space-time boundaries.** It is not necessary to spend long, uninterrupted hours working from home; on the contrary, interrupting to "change the channel" can help to concentrate and restore physical and mental energy. What is important is to respect the entrance and exit gates that physically and mentally separate the contexts. These entrances and exits must have the right timing, such as small breaks of 5–10 minutes or real time slots of 60–90 minutes. What must be avoided at all costs is to remain constantly in the home-work-family, other limbo without fully devoting oneself fully to one or the other.

9. **Constantly monitor people's energy and concentration levels**. This is the most frequent alarm bell, but also the least listened to in working conditions from home. Here people's awareness and self-discipline must be stimulated by the company in all possible ways: communication, training, managerial supervision, and frequent feedback. Distractions in remote work are multiplying and now more than ever we need to spread a highly concentrated work culture that avoids multitasking. And continually stimulate people to work this way through frequent checks and feedback from the field. You can use specific applications to do quick surveys, use the analysis tools of the communication platform you are using, but don't forget to use the most powerful method: increase the frequency of quick informal meetings between manager and employee with this specific objective.

10. **Distribute copies of the book you are reading**!

The New Criteria of (Remote) Leadership and Performance Management

Work from home will no longer be evaluated based on hours worked, but on the basis of results achieved and consistency with operational principles and key habits that the company wants to spread. This means that the company will be forced to specify the objectives to be achieved, declining them by departments and work teams, up to the assignment of targets to individuals, which may partly or totally coincide with the objectives of the department or project team to which they have been assigned. The work of defining and declining targets can be long, but it ensures incredible coherence of action once calibrated and initiated. Is it enough just to evaluate the results achieved? For a company that wants to pursue performance and well-being at the same time, and realize its vision by living its values, it is not only not enough, but it is dangerous to evaluate only results. Alongside the results, the behavior required to realize the corporate vision, maintain value coherence, and build the corporate culture toward which to strive must be monitored.

In Smart Working conditions it can be difficult to outline the behaviors to be observed and evaluated, but perhaps this makes it all more necessary.

Behavior should be easily observable by anyone and should derive from the values you want to see experienced in your company, the vision of the company you want to realize and the results you want to achieve.

The set of behaviors experienced—and observed—will characterize the culture being experienced in practice and not in theory. If I want to change the culture, I will have to work on the specific behaviors and the enabling conditions for those behaviors: skills and competences, surroundings, stimuli, motivations, and beliefs.

If the company, for example, wants to pursue the objective of reducing the overall time spent in meetings—virtual and face-to-face—the behaviors to be defined, shared and made public are all those that contribute to the achievement of the objective and many of these flow directly from what was described in the previous paragraph: compliance with shared rules, convocations with a clear description of objectives and agenda, restriction on the number of participants to the minimum, convocations within agreed time frames.

What is the difference between a written and "imposed" rule and a lived behavior? The behavior experienced must be placed under the continuous observation of colleagues and managers and it is the only element, together with the results, to be considered for quick, specific, and constant feedback, celebrations and various rewards, promotions and—unfortunately, wishing as a remote possibility—dismissals. This is the same difference that passes between the value pictures on the walls of many companies and the values lived in the company itself, which are very often misaligned with what is stated. Whenever promotions, rewards, dismissals, and staff appraisals are not made based on the declared values, you are living values that are different from what is publicly displayed.

The role of the manager in the development of a new culture is crucial because such a figure must ensure the continuous monitoring of behaviors that are functional to corporate strategies. It will be like the coach who constantly monitors the technical gesture of his athlete before even seeing his results. In Smart Working conditions, this concept becomes even more important, because one will have to deal with distance, isolation, and the lack of direct visibility of many behaviors. The manager will have to sharpen his leadership and establish even stronger relationships with his employees to take over what has been said so far and continuously give precise feedback. For example, he will always have to set concrete goals to be achieved and evaluate the ability to achieve them in the *shortest possible time*. Why do I keep insisting on "in the shortest possible time"? Because **time is the main factor that forces you to truly pursue your personal excellence through ever more evolved forms of self-organization**. A good habit to never forget in this period is to set regular one-to-one discussion appointments between managers and employees. You choose the right frequency according to the volume of information and variability: it can range from daily—15 minutes may be enough—to monthly.

Testimony

I work for a defense company that, like all others, had to react quickly to the changes in the pandemic's scenario. I have to say, with some pride, that having decided on smart working, it only took two days for more than 400 people to start working from home with their laptops connected to the IT infrastructure (and this has stood up great!). Although from a technological point of view it worked right from the start, from a managerial point of view I started to have doubts as to whether I could manage my resources remotely while guaranteeing productivity, deadlines, and proper interactions between teams.

I took my notes from the Lean Lifestyle course and quickly found the necessary ingredients: designing new habits, delegation, presidium planning, which is fundamental in this way of working, planning by objectives, establishing regular cadences, meetings with visual management, etc.

Of course, being a remote manager is different, but in the end the bricks are the same, and so in practice I have planned more ways and applied a principle that I have always learnt from the Lean Lifestyle: experiment, measure and then see if it works and adjust the shot.

Here are some of the most fruitful behaviors:

- Give clear objectives at the beginning of the week for each person. I ask them to propose their own and then I maybe adjust them to the expected results (establishing a suitable feed metric).

- Quick and frequent 'online' dating with each person to understand problems, hesitations, uncertainties, especially in the first moments of lockdown (clearly in the slot designated for this).

- Checks at the end of the week on targets, what worked and the problems they encountered, both technical and behavioral, and ranking of progress according to the established metrics.

- Bi-weekly team meetings for exchanging ideas: I realized that second-level interactions (the coffee machine kind, to be clear) between colleagues from different teams, fundamental to elaborate ideas outside the box, were not possible with smart working, so I organized a virtual coffee.

Certainly, at the beginning this meant a little extra effort, but how to manage and recover energy is another gift of the Lean Lifestyle course. Results in hand I can say that it worked very well, so much so that the company will probably maintain a smart working quota even at the end of this health emergency.

—Giulio Cesare Grande, *Head of ELINT & EW*
—Command and Control Product Line Electronics

Design and Deliver of a Specific Training Plan

After an initial evaluation phase, with the data in hand resulting from behavioral observations and objective surveys, it is necessary to construct training and information pathways to support the adoption of behaviors and skills necessary to improve productivity and well-being in all possible forms of smart working: from home, in the office, wherever you work- Work From Anywhere. Training is a powerful weapon at a company's disposal, but it must be used well. It must be designed in such a way as to ensure, first of all, consistency with the objectives to be achieved and with the culture to be spread, and it must be designed with priority given to the gaps of competence that have emerged at an early stage or that emerge later. Another important element in training in the digital era is the concept of a skills training gym, that is designing training in such a way as to have continuity of stimuli and feedback diluted over time, just as happens in a sports training program. Today, this concept is made easier by the possibility of having digital platforms available 24 hours a day, seven days a week, with specific, short, and focused modules lasting no more than 15–20 minutes, so as to ensure the possibility of choosing the content deemed necessary from time to time. As in any self-respecting training program, the plan must always be linked with two elements: the mentor/coach who supervises, provides support and feedback, internal or external to the company, and the link to the results being achieved. If the latter do not change for the better over time, it is very likely that the training and mentoring have been ineffective and perhaps need to be changed or more simply adapted to the real needs of the individuals and companies concerned.

Lean Lifestyle Story: Poste Italiane

Lean at Home and at Work: Managerial Growth According to the Lean Lifestyle Method

> "It has been one of the most rewarding projects of my career, a project of which I am extremely proud because I have always believed that training should have as its ultimate goal to make people feel good in the organization, maximizing the correspondence between the motivations of the individual and the needs of the company."

The speaker is **Luigi Mazzotta**, Head of Human Resources and Organization Training—HR Business Partner Poste Italiane, promoter, together with the Group's Corporate University, of the "Lean Culture" Project, an initiative for innovation in work approach models launched in 2018 in parallel with the industrial transformation project that since the same year has been leading to the implementation of Lean methods, tools and techniques in the Postal Sorting Centers, real realities with an industrial organization.

Objective: to generate a strategic alliance within the company, between people and between functions, and to proliferate Lean principles throughout the organization.

"The intuition came in the autumn of 2017," Mazzotta recounts,

> The strategic decision to start the Lean Transformation of the Sorting Centers had already been made, and during a business trip by train with the colleague who was in charge of the Lean Program at the time, we convinced ourselves that we should accompany the technological change of the production units by encouraging the development of Lean behaviors also in the people not directly involved in the industrial processes. A transversal Lean culture had to be born, made up of a common language and new behaviors, that did not remain only in the technical areas.

Poste Italiane has therefore decided to launch this important managerial and professional growth plan, designing the "Lean Culture" course with the aim of introducing a new approach to work that is functional to criteria of efficiency and effectiveness, but also to the greater well-being of people at work.

The very high level of appreciation for the training program, initially designed according to the Lean Lifestyle methodology, had a particular impact during 2020, as the methods and tools adopted coincided with the generalized start-up phase of the remote work of the people involved, especially those of staff, enabling them not only to achieve constant, indeed often better professional results, but also to verify the positive impact on their individual well-being, in an undoubtedly new and difficult phase in the lives of people, all of whom were seeking a new work–life balance.

The recipients of the first phase were all central and territorial managers in the logistics area and PCL trainers, specialists and heads of operational structures who carry out internal training activities, a fundamental link for the subsequent process of disseminating and implanting knowledge and behaviors functional to the "Lean Culture" in Poste Italiane.

Specific learning objectives have been identified for each of the targets involved and dedicated teaching structures have been defined. As for management, three classroom days including experiential parts related to the contents, with simulated lean-meeting and lean mailing, comparison with external realities of excellence and launch of action plans on the introduction of Lean practices in the managed structure. For PCL Trainers, three classroom days focused on the principles of the Lean model and how to apply them effectively, but also on how to deliver them in a simple and comprehensible way through training. The training of these in-house Trainers included their alignment on the "Lean for Operating Facilities" teaching package, designed to also give PCL operational staff the fundamentals of Lean culture and motivate them to individual responsibility on waste elimination and continuous improvement, for small and large results, but which were continuous and progressive.

After this first phase, the training program has become a consolidated format and has been adopted to train technical staff managers and production managers of the Sorting Centers, "senior" professionals of the central and territorial structures, and production managers in Delivery (first launch of the Lean program also in delivery), for a total, in 2020, of 610 people trained. The program will continue in 2021 with a total target of another 750 people, including department managers, shift managers, and team leaders of the Sorting Centers.

Design and Implementation of an Integrated Communication System

When people are physically distributed, synchronizing communication becomes more difficult. Communication activities, information exchange, brainstorming and problem solving in general, are generally made more complicated. Video conferencing platforms certainly help people to talk to each other at a distance, but without some structured corporate actions, the overall communication can be reduced, can take undesirable directions and, above all, can lead to people becoming estranged, losing a sense of belonging and weakening team spirit. It is therefore essential to prepare a communication plan that includes all the aspects that I am dealing with in this section and that the company has decided to welcome and make its own system elements.

It is necessary to carefully explain everything that will characterize the new way of working and to include all these explanations both within an asynchronous digital system—that is, one that can be used by people when and how they want (such as Microsoft Teams, Slack, Spike, Fleep, Rocket.Chat and Mattermost)—and within synchronous communication sessions carried out directly by company managers and the people involved in live video. Asynchronous systems, unlike the simple e-mail channel (which tends to shrink in the coming years), allow an infinite set of ever-evolving communication functionalities, including the management of channels or groups dedicated to different topics; the instant exchange of information and documents; messaging activities; "call-for-help" activities regarding a problem to be solved; and automatic synchronizations of updates from other corporate platforms. Care not to confuse videoconferencing systems with communication systems corporate: Zoom, probably the most famous video conferencing tool is fine as a synchronous communication platform, for meetings, direct and webinars, but is not designed for various asynchronous forms of communication, as well as for remote collaborative work capabilities.

The presence of a communication platform will help the company to structure the dissemination of its messages, protecting them within the company boundaries, and at the same time promote the free interaction of people, creating the mechanisms, virtual, of the corporate coffee machine. However, this will not be enough to guarantee the full effectiveness of communication. It is important to schedule, with the appropriate frequency, synchronous communication moments at several levels and with specific purposes: one-to-one manager-collaborator, group and company-wide. The purposes of these virtual meetings should include moments of communication of what the company has decided to do in the short and medium to long term, moments of pure socialization, virtual aperitifs or coffees and moments of comparison and collection of bottom-up feedback.

Making integrated communication in a company means building a mosaic composed of dialogue tiles of different natures. And as with any mosaic, only when it is completed can the overall picture you want to bring to the attention of the observer be clearly seen.

Implementation, Monitoring, and Continuous Improvement of Smart Working

The plan described so far must be considered a methodological outline that can be applied at any time during your (real) Smart Working experience.

The implementation plan and its monitoring and review steps must be adapted company by company. One of the most widespread practices in companies with an established Lean mindset is the so-called *hansei*, guided, personal, or group reflection. This practice can be very useful in the review phases of our smart working business plans, and to implement it we just need to master the four key questions of the method:

1) What were our objectives?
2) What have we achieved?
3) What caused the gap between what we wanted to achieve and what we achieved?
4) What countermeasures and corrective actions can we define?

During the *hansei* it is very important to make objective data and facts speak for themselves, to gather input from people operationally involved in the processes or projects we are reviewing and to be forward-looking and pragmatic in defining feasible and reviewable action plans in the next *hansei*.

Taichi Ohno, founding father of Lean Thinking, stated that wherever there is a process there is always the possibility of continuous improvement, but this means above all that one must have defined a process before going to action, otherwise it becomes difficult to understand where, what and why to improve. In fact, Ohno himself stated that there is *no Kaizen without Standards*, that is, only what has been made a standard, visible, objective, and shared, can be continuously improved.[1] I would add that **you only improve what you can measure**. So, if you want to aim to improve your results and those of your company, always ask yourself which processes, what behaviors and what measures to adopt and monitor. Also, and especially, in Smart Working conditions.

Lean Lifestyle Story: Streparava

The Lean Lifestyle Company Also Online

400 to 5. The challenge of **Raffaella Bianchi**, Human Capital Manager, is to make the work of the Streparava Group's personal office agile, effective, and stress-free. With her 4 colleagues, she must manage the needs of around

400 employees. A continuous and intense flow of seemingly unmanageable activities without interruptions and in a smooth way.

"Our way of working is changing radically thanks to the Lean Lifestyle," says Raffaella Bianchi, "The COVID-19 emergency has accelerated the process of improvement in the organization of our work by leveraging the digitization of processes and information. We have made it a specific improvement project that, in short, wants to make employees less dependent to make our work smoother and at the same time give a more punctual, efficient, and safe service to our colleagues."

The structured activities implemented were as follows:

- Employees can get in touch with the personnel office by setting up online appointments with slots lasting 15 minutes each.
- Without booking, the employee can only go to the personnel office during specific time slots.
- Staff communications are now 80% online and during the course of this year, the factory notice boards will be completely removed.
- A WhatsApp business group was implemented to timely inform employees, especially those who do not yet have a personal e-mail.
- The sharing of FAQs, documents, and information for any needs (such as holiday requests, information on rights) on the company's digital platform. Today for all employees, and during 2021 for all workers.
- Launch of an internal communication platform for all employees.
- Training, support, and opening of info-points in the company to facilitate the transition from digital communication.

Raffaella concludes:

In March 2020, we found ourselves working from home overnight. There was a strong risk of losing the connection with colleagues and the company, but above all to be overwhelmed by fear and the amount of unforeseen but necessary work, such as managing buffers, to ensure the safety of all employees. But we were ready and the only certainty we had was our work, the organization we had given ourselves. In a world that was not normal, the work was normal! Only the mode had changed.

How to work and what to do were the certain things at an uncertain moment. This gave us a huge well-being.

Lean Lifestyle Story: Siemens Italia

Smart Working, the Real Thing

If the coronavirus emergency has led remote work to become, by necessity, a constant in Italy, there are companies, such as Siemens Italia, that have believed in the principles of Smart Working, the real one, since 2011 when the first experiments were started. After several waves of extension to ever-wider swathes of the corporate population, the Smart Working model was adopted as the definitive measure. In June 2017, the collective agreement with the trade union was signed. In the pre-Covid-19 period, each smart worker, who was not engaged in the production of physical products, worked outside the company about 1–2 days a week.

> About a week before the first governmental lockdown was decreed, three phone calls over the weekend coordinated by the newly formed Crisis Management Team at Siemens Italy were enough to prepare us for the new conditions and decide on "full smart working." We were already ready.
>
> **Ermanno Delogu**, CEO in KUKA

And it is not a matter of mere remote work, but of the methodical application of a set of conditions that are gradually changing the way of working in Siemens Italia. This is what **Eirini Nektaria Karamani**, Country HR Head at Siemens Italy:

> The first and greatest change we are making is related to the mindset of people: from activity orientation to hours worked, to a focus on the results. And this represented and represents the biggest challenge. It's true, we had to adapt our offices to make them conform to a more flexible way of working, without fixed workstations, but bookable from time to time, we have had to integrate our IT infrastructure to make remote collaboration of work groups possible and effective, but the biggest barriers remain those related to one's own convictions on what is right and what is not right to do in the workplace.

Today, years after the first trials, no one complains if you are not in the office or at home, because they know that you can work wherever you see fit. And this is a real revolution, especially in Italy, which proves that the concepts of *WFA—work from anywhere—*seen in Chapter 11 are not part of an overseas world alien to us. Working from home is just one of the options. A revolution that today is only just beginning, according to Eirini:

> It is already possible today, although the practical options are not yet so abundant, but I think that in the future it will be normal to be able to work in a café, in a park equipped with Wi-Fi connections, in a co-working center, and even to choose the city in which you prefer to live and then work. This will bring radical changes compared to today, because the workplace as we have always interpreted it today will be replaced by more and more flexible forms of voluntary work, true mobile working. Perhaps the big cities will empty in favor of a decentralization of workers. We will be able to hire people who do not reside in the city where the company has its headquarters, widening the range of possible options for both the company and the people.

If this is what you see on the surface, what is needed at the base to realize a seemingly utopian model is very deep. Siemens Italia represents the flagship and pilot site of an experiment that the German multinational wants to pursue worldwide. One of the company's strategic guidelines is that of the "Growth Mindset," or the attitude to continuous growth that is being cultivated in all people. "Learn something new every day" is the new mantra according to Eirini to be adopted by everyone in every situation. Evaluation systems that assess the achievement of objectives once a year are "obsolete" in this new context because it is unthinkable that what I have envisaged today will remain unchanged 12 to 15 months later, with the speed of the changes taking place. You need "organizational agility" and a "culture of adaptation" to realize the model of individual and corporate excellence that you have in mind.

> "It is important for managers to be able to link results and employee growth. That is why it is necessary to recognize successes as they happen and to reward people in good time, just as it is essential to be able to 'correct the shot' in real time and to accompany people to continuous improvement, not just once a year," says Eirini, "so that everyone's focus is definitively shifted to continuous growth and individual excellence as a permanent

process. We have already disconnected the performance management system from the bonus system. Now we are transforming our Performance Management Process into continuous growth conversations. Once salaries are aligned with the right market values, people must focus on growth, the future and innovation."

Another essential keyword to support the entire Smart Working concept at Siemens Italy is trust.

We are reinforcing a "trust-based environment," that is a working environment based on mutual trust, a condition without which it is impossible to achieve a transformation that is first cultural and then infrastructural. In a climate of trust, aligning with managers to share priorities must become a daily practice, rather than just including numerical targets into a cold evaluation system.

The benefits achieved in these years are in line with those already described in Chapter 11: increased productivity, reduced fixed costs, and reduced physical space management. But the greatest benefit being achieved relates to the cultural transformation taking place. A transformation that will enable the company to keep moving with great ability to adapt quickly to change.

Siemens Italia's plans include reviewing the workplace in an agile way and making all the necessary fine-tunings to be ready, after the Covid-19 crisis, to offer people the WFA mode—*work from anywhere*—leaving people free to choose but preparing the entire system for hybrid and collaborative work modes on a permanent basis.

"It is not easy, but people are happy with these changes. I foresee that by 2023," Eirini concludes, "when the pandemic events are over, it will be normal at Siemens Italy to work 2–4 days a week outside the office, wherever you prefer."

Note

1. J. Liker, *Toyota Way. 14 Management Principles from the World's Greatest Manufacturer*, McGraw-Hill, 2004

Chapter 13

Visionary Companies Capable of Turning Strategy into Action

Everything we have seen so far in this book has proven to have a concrete impact on the way thousands of entrepreneurs and managers work toward the direction of simultaneous increases in performance and well-being. But if we want the change to extend to the whole organization in a way that is synergetic and complementary to the other initiatives underway, we must necessarily develop a single, multi-year strategic plan for corporate evolution.

We no longer live in the era of magical industrial plans and long-term "business plans" because they become obsolete at the very moment when you painstakingly write and formalized them.

With all due respect to control-obsessed managers, investors, and entrepreneurs clinging to old traditional logics: data, forecasts, detailed targets, multi-year action plans, and cost and profit item analyses that split the hair in four are no longer needed.

Avoid wasting your time. Today, we must learn to combine forward-looking vision with a healthy, structured pragmatism, and we must equip ourselves with appropriate business systems to do so. We must know how to drive the company like a car, along uncertain roads, never driven before, and in thick fog. We must know where we want to go, but it is vital to proceed very carefully, meter by meter, and be ready to dodge dangers, slow down, and speed up depending on what we encounter along the way. In business terms, we can say today that it has become vital to transmit to the entire organization a medium-long-term strategic direction and at the same time an extreme focus on short-term continuous improvement.

DOI: 10.4324/9781003474852-13

Unfortunately, many mistakes are made in many companies in terms of strategic planning and balancing short- and long-term demands. I have summarized the most common mistakes I have observed in recent years. There are ten of them. You will probably think of others, but I believe that avoiding these can already take you a long way forward, should you decide to take on the challenge of creating your new strategic business plan according to Lean Lifestyle principles.

The Ten Mistakes Not to Make in Strategic Business Planning

1. Instinctive decisions and wrong target selection
   This is the mistake we encounter when decisions are made with little data and partial information. When such decisions largely reflect the desires, patterns, and mental schemes of a few people instead of reflecting objective interpretations of risks and opportunities for the company. We are conditioned by our personal beliefs at every juncture of our lives, so much so that we often tend to confuse instinctive behaviors—our intuition—with behaviors derived, more simply, from what we are convinced of. While this is normal, it is not in the context of long-term business decisions based only on our convictions and instincts. An effective strategic planning system must help the entire organization avoid this dangerous cognitive bias. **A clever and effective way to broaden our own mental schemes is to integrate them with those of others**. Therefore, strategic planning must be conceived as a moment of true team building, where the result reflects a work of integration, mutual calibration, and constructive discussion in the interest of the whole company and not of the single function.
2. Absence of clear basic elements: values, vision, and mission
   When the business strategy has no connection to an ideal long-term vision, but is based only on opportunistic and speculative tactics, the probability of making the right choices will be very low. In any company, people are motivated by the deeper meanings and motivations of their actions. **The stronger the link between what you do and why you do it**, **the more likely you are to have motivated and proud people on board.**

3. Poor communication process to team and stakeholders

   When a strategy is not structured and well communicated to everyone, it is no longer a strategy but a set of high-level statements and objectives that remains only in the cockpit and does not fulfil one of its main tasks: to "bring on board" all the people. Strategy in the company should explain what direction one wants to take and why so that the efforts and contributions of all people converge in the same direction. To be an effective communication, it must also be conceived as bidirectional, that is to accept and include feedback and calibrations of content along the way. To avoid making this mistake, it is therefore crucial in strategic business planning to design, and execute, the communication process to support it.

4. Unidirectional and authoritarian process

   We are all attracted by the charisma of successful entrepreneurs and managers who have built great business empires and emerge in the collective imagination as heroic figures and leaders of other times. And we remain enchanted by them. Illustrious names from overseas or blazoned names from glorious Italian family businesses. But what shines at first glance often shows its true face up close—for better or worse. You will almost never find a lonely, authoritarian leader in charge but solid corporate management systems and a large team of people who are inspired and guided by the figure of the leader who, however, does not replace them in the decision-making and operational running of the company. The difference is subtle but fundamental: **making employees feel that they are substantial owners and conductors of the company instead of mere employees and executors of others' ideas.**

5. Lack of monitoring and progress monitoring

   In the next pages, we will see that the strategic planning of a company can be divided into three phases: genesis of the strategic directions to be undertaken (strategy generation); operational declination at all levels of strategy (strategy deployment); and implementation, monitoring, and operational governance of the strategic plan (strategy execution). Often a company's strategy stops at the first phase, not devoting the right amount of space to the planning and implementation of the next two phases, which are by far more important if we look at it from the point of view of those who work in the company every day. When and how to monitor, how to react to deviations, how to involve people, how to value progress, how to link all the elements of the strategic plan together are examples of the execution phase of a strategic plan during which the company must decide whether to persevere on its course and whether to calibrate or change direction. Realizing that you have taken the wrong path when you arrived where you did not want to has always been a cause of frustration, unnecessary waste of energy and resources, but with the current speed of contextual changes, it can cost us dearly compared to the past.

6. Poor management of time constraints and resource allocation
   Companies often make the same mistake that I find in individuals: planning with resource overload. Workloads are not managed at "finite capacity" but at "infinite capacity," as if people had days and weeks of unlimited hours at their disposal. We all make this mistake, but the most serious thing is to make it an organizational behavior. Not managing time constraints and the allocation of available resources leads companies to fail to recognize capacity shortages. Such deficiencies are only recognized when the problem erupts in the implementation phase. The problem must be solved by managing the number of projects that are launched at the same time, by adjusting the compositions of project teams according to the real overlaps between one project and another, and by resorting to agile and fluid forms of organization, where small, autonomous teams with definite resources are focused on very few projects in full immersion mode.

7. Planned activities without sufficient attention to failure factors
   Everything we plan often does not happen exactly as we imagined. Sometimes some factors will contribute positively to achieving the desired results, and sometimes other factors will contribute negatively. In the context of planning and executing a winning business strategy, one of the most important skills to be developed at managerial level is that of preventive care of possible failure factors. We could more correctly define it "risk management" if we wanted to use structured methods for risk management in any project, although it is often sufficient to ensure that we place a few key questions in a structured manner at the right time in the strategic planning process and make the answers compulsory. What constraints, internal or external, do we have to achieve these results? What skills are needed? What external factors can influence our plan or each of our strategic projects? What business support systems or tools do we need to increase our chances of success?

8. Self-referentiality

   Often many companies do not grow as much as they could because, paradoxically, they do not plan to do so and tend to repeat strategic moves made in the past, because, after all, they give more security. Growth-oriented strategic planning looks much more outward than inward, speaks much more of the language of the market than the company's historical language, and tends to question what has been done so far, even when it has brought good results in the past. In the context of extreme variability and rapid changes in market needs, a common mistake is to focus on competitive factors defined from the perspective of the company and not of possible buyers. This mistake also stems from limiting observation to those who have already bought from us in the past. If we want to break down the restrictions of self-referentiality in our business strategy, we need to see our customers, past and present, with different eyes and ask ourselves what we can do to make them buy from us again, maybe expanding our supply perimeter and begin to study more those who today consciously choose not to buy from us but decide to buy goods and services from other competitors. Finally, let us train ourselves to observe much more those who do not even know our company. If our strategy does not include these elements, we are very likely to remain in our comfortable zone of self-referentiality.

9. Poor balance between incremental and high-impact innovation

   A good strategic plan should protect the company from risks related to the obsolescence of the business model, technologies, products, and services, as well as help to prevent crisis situations due to new regulatory and legislative constraints, new market conditions, and anything else that undermines profitable business operations. At the same time, a good strategy should help the company to seize new market, technology, product, and service opportunities to grow profits and margins. This is achieved by balancing incremental improvement initiatives of what already exists and initiatives aimed at creating new market spaces or aimed at creating from scratch competitive factors that can make a difference to competitors: high-impact innovations. When there is too much propensity for the new, one makes the mistake of not maximizing the returns on what a company has built up over time, for example, through the pursuit of greater efficiencies, perimeter expansions, or expansion into new geographical markets. When there is too much propensity to exist, it leads to the mistake of not looking at new opportunities or risks of obsolescence in one's business.

10. Luck syndrome

The typical mistake in situations where a company's growth is linked to a sector that is growing and not to a specific strategy or when growth is linked to a strategic move that was successful in the past but not followed by subsequent winning strategic moves. In these cases, the illusion of continuous growth is dangerous, because the levers of growth are out of the company's control and do not show dangerous structural deficiencies until the problem bursts into all its urgency. We could also decline the luck syndrome into its opposite: bad luck syndrome, that is finding oneself in an industry in rapid decline or with profound structural changes and resigning oneself to corporate extinction. Here again, the mistake is not to focus enough on the levers that the company can act upon instead of allowing itself to be inebriated or depressed only by external and contingent factors.

Simplify and Master Corporate Strategy: Hoshin Kanri Management

Unfortunately, in 25 years of work, I have observed many companies not aligned toward their strategic objectives, and even when I have encountered a formalized strategic plan, very few times have I seen it linked to elements shared and understood by the entire company population. Most of the time, however, the problem was not the "bad intentions" of the entrepreneurs or managers but the absence of the right corporate system to implement foolproof planning. What I propose to you as a corporate system for managing your company's strategy is the evolution of a well-known method, called Hoshin Kanri, born in Japan and applied by hundreds of companies worldwide since the 1960s.

With the Hoshin Kanri, you can set up a strategic planning that allows you to transfer and share corporate objectives to all levels of the organization, transforming them into operational projects and punctual actions, reducing waste due to both the lack of clarity on the common direction to be taken and the absence of a solid strategy management process. This methodology has its roots in other earlier management practices: the management by objectives (MBO), the method theorized by Peter Drucker in 1954, and plan do check act (PDCA), the method developed by Edward Deming in the 1950s in Japan.

Operationally, the Hoshin Kanri is one of the most powerful Lean methodologies for the company because it ensures that a company's strategy is correctly created and successfully executed by the whole organization, thanks to the fact that projects and initiatives are shared and chosen from time to time to achieve business results.

In-Depth Analysis

The three keys to Hoshin Kanri's success are:

■ the application of PDCA to a company's strategy, with precise periodic process phases divided into planning, execution, control, and definition of countermeasures.

■ the application of the so-called catchball, or the systematic "top-down" transfer of targets to be achieved, while leaving each transfer step free to calibrate and negotiate the targets.

■ the empowerment and delegation to the operational teams of the definition of how to achieve the set objectives.

In this way, the Hoshin Kanri becomes the single strategic plan for the entire company that drives focused progress and improvement actions at all organizational levels of the company. They all point in the same direction, with a clear common long-term direction, but focusing on short-term progress to be achieved.

One of the elements of organizational success is communicating the objectives and expected results in a way that is relevant to everyone. High-level objectives are always declined into lower-level objectives to obtain consensus from everyone. Providing structured bottom-up feedback on decisions that impact a particular business area, or work group, increases involvement, cohesion, and alignment, reducing friction and misunderstandings within the organization.

How does the "catchball" process work in practice? Imagine that management decided and created a strategic objective to be achieved. The next step is to ask the organizational level below for analysis, feedback on feasibility, possible blocking points and constraints, and above all, ideas and projects that can be put into practice to concretely achieve that goal. After negotiating, sharing, and achieving convergence between the two organizational levels in comparison, the "catchball" process proceeds until it touches the most operational level of the organization. The ball passes from hand to hand from top to bottom and vice versa, and everyone contributes with their ideas, embracing and solving specific problems, needs, and constraints that may emerge in this process.

Like any other Lean technique, it can take years of practice for a strategic plan to succeed. It is an organizational learning journey that requires a truly collaborative culture in which everyone invests time and attention in each other's success.

These are the seven operating steps of the logical process of the Hoshin Kanri:

1. Definition of vision, mission, and values (winning aspiration): why this plan?
2. Long-term strategic goals (3–5 years): what do I want to achieve over time?
3. Short-term strategic objectives (12 months): what do I want to achieve first?
4. Choice and definition of strategic projects: how to achieve this?
5. Defining of metrics and KPIs to be improved: how much and when?
6. Allocation of resources: who will do what?
7. Monthly and annual review plan with deviation analysis and countermeasures.

Winning Aspiration

The first phase of strategic planning according to the criteria of Hoshin Kanri is underestimated by most companies, and in the best of cases is part of an isolated, one-off job, not closely connected to annual strategic plans. There is, however, a considerable detail to keep in mind for every type of organization. As I have already written, people are not motivated by a company's procedures and processes. They are motivated neither by the procedures and processes of a company nor by the six-month or one-year targets that it sets itself: they are driven to action by the deep motivations that managers and entrepreneurs manage to unleash in them. This should be the main commitment of managers and entrepreneurs, and, in my opinion, this is the managerial commitment that must always precede all others. As Simon Sinek argues, it is important that every company always starts from the why, before focusing on the how, and then on the "what to do."[1]

In relation to strategic planning of a company, this means focusing on what constitutes the "winning aspiration" of an organization, before thinking about long- and short-term goals.

The term "winning aspiration" was coined by Alan George Lafley, one of the world's highest-paid CEOs, who led Procter & Gamble from 2000 to 2010 and from 2013 to 2015, as well as being a board member of General Electric and Dell. Under Lafley's leadership, P&G has become the world's second largest consumer goods manufacturer after Nestlé. One of the reasons for his success was his propensity to make decisions and strategic choices in close connection with the definition of the company's ideal aspiration. P&G was rewritten by Lafley[2] and his team without half-measures in this way: "to be and be recognized as the best consumer products and services company in the world." According to Lafley, in business you play to win and not to participate, but winning is difficult if you don't dare to make difficult choices consistent with your winning aspiration.

Using other terms, Salim Ismail, co-founder and executive director of Singularity University, speaks of Massive Transformative Purpose[3] (MTP) as the reason for the existence of so-called exponential growth technology companies, those that have grown faster than all the others and have been able to answer the following questions more effectively than others:

Why does this organization exist and why does this work?
What problem do we want to solve and how do we want to solve it?
What will be the global impact of our organization?

According to Ismail, fast-growing, exponential organizations always think big. They always want to transform their industry or their community or even society. If a company thinks small, then it is likely that it will not adopt a business strategy capable of achieving rapid growth. The "position statements" of an MTP are aimed at defining the expected identity, not the real one, such as "organizing all the world's information" in the case of Google. They generally do not describe what the company does, but the focus is on what it aspires to achieve. An MTP aims to capture the hearts and minds of those inside, as well as outside, the organization and is the ultimate aspiration and source of inspiration for the entire company.

The Envisioning Process

In the reality of Italian companies, what happens about ideal aspirations and perspective visions? In many organizations, it is believed that the task of organizational "inspiration" is considered with the exercise of writing down the company's *vision* and *mission* is exhausted, many times with the help of consulting firms or business coaches. I have often seen pretty little pictures hanging on office walls or beautiful phrases written on company websites; however, what I have rarely encountered is a structured link of those phrases with the actual strategy and culture of the company. People hardly "felt they belonged." I have often found phrases that had been developed by a small group of people, then communicated to everyone, but were practically detached from reality.

> There's nothing more demoralizing than a leader who can't clearly articulate why we're doing what we're doing.
>
> **Jim Kouzes and Barry Posner**

Numerous studies show that successful managers manifest their dedication to mission and values in everything they do: in their actions, in their dreams, in the example they give, and certainly in their speeches.[4] Even if the goal is not to change the fate of the whole world, every company that wants to reach its maximum human and organizational potential should strive to give profound meaning to its very existence. When a set of shared values and a purpose that goes beyond mere profit is lived out daily, the spiritual energy of people at work is nourished, the energy that comes from the deep meaning recognized in work activities. And when this happens, people find the enthusiasm and motivation that pushes them to face their daily work with passion and not with a mere "sense of duty." There is therefore a need for a shared generation of organizational *winning aspirations*, which must start from the people and then converge toward the company. If we talk about corporate vision, for example, we should start from the aspirations of individuals, from their projections into the future, and then reduce them into the future identity of the company they work for. It is as if the identity of a company were composed of the sum of the different identities that flow into it. Over the past decade, with my Lenovys team, I have worked alongside numerous companies to develop their multi-year strategy. Almost always the initial trigger to the strategy was represented by the generation of winning aspiration through a process that starts from people and then converges on the company. This process, developed over the years, is called envisioning.

In this process, the real "hard core" of the company is explored, defined, and formalized, passing through the emotional and rational grids of its people. The beating heart of the company becomes the *vision, mission,* and *values* generated through a decoding of people's aspirations and beliefs and therefore, much more "felt" than what can be derived from the traditional definition process.

- The *vision* represents the future identity to aspire to: the dream of people and the company.
- The *mission* makes clear why we do the things we do and why we are different: the purpose.
- The *values* tell how we want to work together: the compass.

When these three elements are "aligned" and subsequently linked to the corporate strategy and culture, people are cohesive, making decisions, and acting in one direction, that of choice. As can be seen in Figure 13.1, the *vision* will be the ideal destination of the corporate identity, the *mission* will be the representation of corporate identity, and the *values* will be the behavioral guide even in the toughest moment, supporting the dreams and hopes of the people in the company.

> You've got to think about big things while you are doing small things, so that all the small things go in the right direction.
>
> **Alvin Tofler**

Reinterpreted in this way, *vision*, *mission*, and *values* must be not only simple and ambitious but also easily remembered by everyone: in this way they will become the real daily reference for managers and employees. Only if they are concrete elements for employees, can they really be effective. Regardless of the size of their group, great leaders know that their first responsibility is not about operating margins or procedures but about translating corporate *vision* from a virtual dimension into a set of daily priorities for their employees.

Deoflor followed Hoshin Kanri and declined vision and mission:

Vision: "Innovative, Safe, Sustainable. We take care of people and environments bringing pleasure and well-being into everyday life."

Mission: "With the passion and expertise of our team, we develop and improve technologies and processes to offer Total Daily Care solutions with beautiful, effective, safe, competitive, and sustainable products. We work together with our customers in creating Value for the end Consumer."

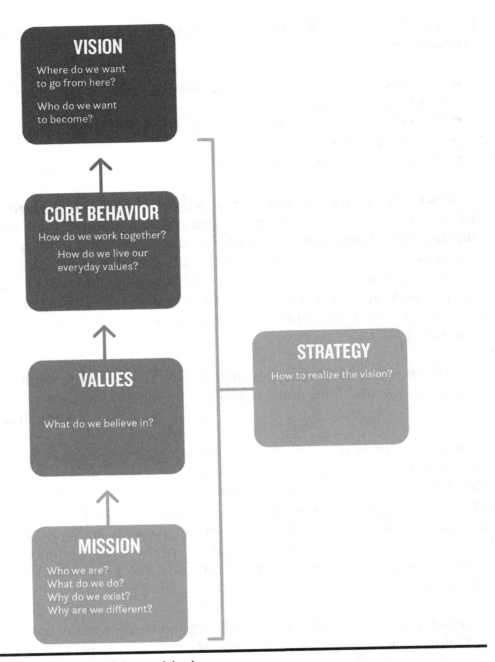

Figure 13.1 The steps of the envisioning process.

The Hoshin Kanri marked a turning point for Deoflor. It is a governance system that has brought on board all the people involved in the strategic projects we have chosen through a shared process. It is a "beacon" for which we all feel responsible; the strategy is not dropped from above, everyone knows how to assess whether we are deviating from it in the course of work; the rigorous project management process that results, through structured steps and the use of A3 management, made our work easier.

Gian Luca Guenzi, CEO Deoflor

Whenever the envisioning process is completed and linked to strategic planning, the consequences (observed over the years) are:

- Increased enthusiasm for a purpose that people have helped to define and convergence toward a precise image of the common "dream" to be achieved.
- Increased team energy resulting from focus and clarity of direction.
- Increased speed in making the right operational and strategic decisions.
- Questioning the current state, when not adequate to achieve the vision.
- Increased motivation to cope better with challenges and moments of stress or crisis.
- Reduction of inconsistencies or loss of focus in the achievement of key objectives.

Streparava also followed the Hoshin Kanri, thus declining its vision: "Passionate, inspired, sustainable. A global professional team recognized as a key innovative solutions provider for the mobility of the future."

It was an exciting journey because Lenovys took our energy, individual dreams of people, as each of us wanted to be happy, and accompanied us in the strategic design of the company. It has been a powerful journey that has led us to define a Vision and a Mission consistent with our feeling, our daily living of the company, then declined with a Strategy Deployment made of specific objectives and projects defined with the full involvement of all.

Paolo Streparava, CEO Streparava SpA

From Management to Travel Rules

I do not see the company in a static way but as a complex machine in constant motion, grappling with a journey full of dangers and opportunities. The direction taken in this journey is represented by the *vision* that is defined. For this reason, it is crucial that this direction is generated, shared, and deeply felt by the people who drive the corporate machine. But as with any journey, we can decide to move from lonely explorers or with a cheerful company of friends, independently or within an already organized context. We can decide whether to use one means of transport or the other, more comfortably, and quickly. We can decide which clothes to travel in, sporty or elegant. Bringing the travel metaphor into our companies, these decisions will represent the key values and behaviors we want to live in the organization. These are the elements that characterize people's choices and interactions.

What are the values of your company?
Can they motivate or influence your behavior and that of your colleagues?
Are they easily forgotten or are they authentic and hit the mark?
Are they spread consistently and effectively every single working day?

Beyond the proclamations and what is written and communicated in brochures, websites, and other institutional channels, the behaviors that are observed daily represent the operating culture of a company. These behaviors reflect the values experienced by people, or the deepest convictions we have rooted, the ones we believe in the most. If, for example, I believe in the value of teamwork, I will try to collaborate with my colleagues in a constructive manner, I will be available when I receive requests for help, and I will not remain isolated to solve the problems I encounter. Defining in a corporate context the values that the group decides to share and to translate into acted behavior is a fundamental step for any type of organization.

Beware of inconsistencies between what is said or written in relation to corporate values and what is done in practice. It is important to pay attention to the behavior of the top and management group, which is the most evident manifestation of the company's values. Every day this group of people is observed and taken as a reference, not so much for what is said, but for how it is said, for the choices made and the actions performed. Entrepreneurs and managers can inspire and engage, or they can demotivate and alienate, much more than they imagine.

There is no shortage of sensational cases in which famous companies have loudly declared certain values such as ethics and integrity, only to have their top executives investigated for aggravated fraud, tax evasion, and much more. The Enron case of 2001 is an extreme one, a US multinational that went bankrupt due to fraudulent bankruptcy from the height of its $130 billion turnover, burnt in a few months to the amazement of the whole world.

It is good to remember that workers are much more loyal, and have much more trust in, to "bosses" who show integrity and are consistent with what is stated.

I also do not believe in perfection and the possibility of absolute consistency. We are human beings, and it will always happen that we adopt behavior that for a thousand reasons may deviate from what we deeply believe in. It happens even to the best parents, in certain moments of stress and tension, to react abruptly and treat their children with little delicacy. And vice versa. But underlying intentions and love will bring parents and children closer, clarify and embrace again. Dialogue, confrontation, and awareness of deep intentions are the basis for building solid relationships even between people in the company.

How to Minimize the Inconsistencies

To minimize the inconsistencies between "prescribed" values and values experienced by people in the company, it is important that three types of structured, upstream–downstream activities are done very well: generation, dissemination, and feedback.

Generation

It is important that corporate values are defined through a process of exploration, sharing, and choice, focusing on those that unite the individuals participating in these work sessions. This is why I always suggest widening the participants to these activities beyond the narrow top management ranks, to listen to different voices coming from the "field." If different values are present in the company, it is good to face diversity at this stage and find constructive convergences, rather than letting inconsistencies emerge in everyday life.

Dissemination

In order to link values to the behavior experienced in the company, it is necessary to design the entire communication process, paying attention to the operational dimension of the values and the impact they will have in practice. Each declared value will have to be accompanied by precise organizational consequences, so as not to fall into a sterile exercise that will not involve anyone. If we want to communicate and spread the corporate value of well-being, we will have to communicate the relevant concrete initiatives to support this (for example, training, psychological support, coaching, company gym and other sports initiatives, and dissemination of good food practices).

Feedback

This activity is probably the most important but often neglected. It involves the design of timely feedback systems to support behavior consistent with corporate values. The first feedback system is the most powerful one, seemingly simpler and more superficial but complex and deep to a more careful analysis: these are the informal feedback that managers give to their employees whenever they interact with them. This feedback cannot be left to chance or arbitrariness of the individual manager but must involve the observation of defined key behaviors. At each interaction, we have a great opportunity to observe how our employees act and provide feedback. Small, quick, and incisive feedback is the driving force behind the calibration of human behavior in the company. If, for example, we want to live the value of teamwork, we have dozens of behaviors every day that will be oriented toward this value: e-mails, meetings, and interactions will give us many opportunities for evaluation. As managers, we should sharpen our capacity for observation and accept our task as regulators of values, through reinforcing feedback for the behaviors we want to consolidate and "extinguishing" those we want to minimize.

The more formal feedback method, the one that should "close the circle," is the corporate evaluation system. If we want to use the appraisal system to consolidate and shape the corporate culture, as I have already written in Chapter 12, we will have to measure both the performance and results achieved as well as the behavior acted out by people in respect to corporate values. In this case, be prepared to carefully face those who are performing but averse to the corporate culture. At the same time, be prepared to value and grow people in the organization who are beneficial to the corporate values you want to live.

The feedback phase can include celebrations and recognitions, periodic events, and various initiatives that aim to keep high the focus on the behavior I want to see acted upon in the company. Going back to the previous example of teamwork, we can each month publicly celebrate the team that has best lived this value. We don't need financial incentives, big celebrations, or big metrics. Just design simple, focused organizational routines.

Without this feedback work, the risk of seeing the first two phases, generation and dissemination, fall into the void is high—almost certain.

Adopt the Farmer Mindset

To do well this patient work of generation, dissemination, and feedback in company, it is important to adopt the mindset of the farmer, instead of the hunter. The farmer does not care about having a result today at the expense of good harvests over time. The farmer knows that the right process leads to the right results. He knows that nurturing the soil, sowing seeds, tending plants as they grow, weeding, tending to sustainability, and rotating crops are all important actions in an overall process to achieve the expected results not only for the current season but also for future ones.

Make the Right Strategic Choices

Long-Term Strategic Goals

After having focused on one's "winning aspiration," in the form of the *vision, mission,* and *values* just reviewed, it is important to set it out in clear strategic directions and "break" objectives with respect to the current situation. This is not a matter of entering forecasts from a typical business plan but of making the right strategic, measurable, and tangible choices that will guide the company's growth over the next 3–5 years. Examples:

- Expand into a specific foreign market, never touched before.
- Develop and market a new product line.
- Adopt a new technology in the company.
- Acquire to enter a new market or acquire new strategic skills.

This is the central part of the whole strategic plan, the one in which the ideal aspirations of entrepreneurs and managers result in business opportunities that can either take the company in precise directions on its journey or address in a structured way situations that put at risk the very survival of the company.

Knowing how to base a strategy means making winning choices and how to implement them diligently.

Many companies encounter problems in making bold choices, even before they have the most effective system for their realization. Many fronts are pursued simultaneously, many strategic directions and many product lines, with the illusion of having more cards on the table and thus a better chance of ultimate victory. Unfortunately, in most cases it does not work this way, unless one has built up over time a conglomerate composed of numerous sub-companies with a high degree of autonomy. For example, in the case of a multinational giant such as P&G, with over $70 billion in turnover and around 90,000 employees worldwide, the organizational structure is divided into a series of divisions with a high degree of autonomy, to the extent that they are comparable to independent companies with several common infrastructures and services. But once inside the single division, one inevitably discovers that the energies and capacities of any company are limited. The experience of all organizations shows that focus rewards much more than generalist dispersion. Growth is often more sustainably achieved by vertical rather than horizontal expansion, that is, with a lot of business on a few fronts rather than with little business on many fronts. How do you improve your ability to make good and courageous choices?

It is important to distinguish between two synergistic fronts. Using the scheme in Figure 13.2 dear to my friend Filippo Passerini, former CIO and Global Business Services P&G, each company must choose the playing fields in which to compete and the strategies to win games in the selected playing fields: where to compete and how to win.

1. Which markets to address?
2. Which market problems to choose?
3. What solutions to propose?
4. With which product lines?

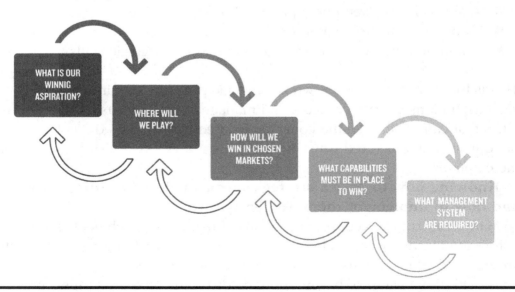

Figure 13.2 Scheme for choosing the "playing fields" in which to compete.

5. With what distinctive features and competitive advantages?
6. Which industrial, technological, and organizational levers should I use?
7. What business management systems will I have to adopt to support the choices made?

These questions should be asked in the sequence presented to stimulate the right strategic reflections. All too often I have noticed the outdated approach whereby companies focus for a long time on one or more developed products and then wonder how to promote, sell, and market them. Self-referentiality. Starting from the markets, on the other hand, and from the specific problems of those markets, will bring companies on the *genba*, during people and real problems, and will recall solutions that lead to coverage strategy skills and high-impact innovation. It will sharpen the ability to make the right choices and generate value.

In this phase of bold choices, the company has the great opportunity to make tangible its attention to the impact of its actions on people, society, and the environment.

Long-term objectives, typically 3–5 years, in addition to being related to the winning aspiration, must represent in synthesis the answers to the questions just seen, to clarify the strategic guidelines are to follow in all the key corporate areas: innovation, technologies, leadership model and corporate culture, organization, key processes to be improved, key competencies to be developed. The objectives defined in this phase will guide the subsequent phases in strategic planning: annual objectives and strategic projects to be launched.

Annual Strategic Objectives

The first operating principle of a company that wants to become a Lean Lifestyle Company is that of vision and execution. We want to create a company that encourages people to think, dream and realize big. Dream big and act small every day. That is why it is important to have a monitoring system that evaluates in the short term (coming months) progress toward the long-term goals (3–5 years). The first step is to break down long-term goals into short-term objectives, usually with a one-year horizon. In this phase of strategic planning, a set of quantitative and measurable targets are defined that "pull" all the annual improvement activities of the company.

If, for example, the long-term goal was to expand abroad, the targets to be achieved in the next 12 months must be precisely defined at this stage: which country, which products, which turnover target. Next year's targets can be included at this stage, which are considered preparatory to the achievement of the overall set of long-term goals.

From Strategy Generation to Perfect Execution

90% of managers and executives admit that they have not successfully implemented part of their strategy, causing waste of resources, destroying corporate productivity, and risking a misalignment of people with strategic objectives. One million dollars is wasted every 20 seconds worldwide due to poor corporate strategy execution. That's two billion dollars every year—equivalent to the GDP of Brazil. These are the embarrassing data that emerge from an authoritative 2017 research conducted by The Economist Intelligence Unit (EIU) on over 500 senior executives of large companies globally.[5]

Any good strategic choice, if not made correctly, causes considerable damage both financially and in terms of organizational and moral cohesion of people. Unfortunately, the same research shows that **63% of the managers surveyed admit that implementation is not seen as a strategic task**, despite its crucial contribution to the company's success.

For an excellent strategy implementation phase, the following steps must be taken with great care: definition of strategic projects and key metrics, resource allocation, governance system and monitoring. Let us analyze these phases in more detail.

Definition of Strategic Projects

This phase of planning is called by Anglo-Saxon colleagues as "deployment" and is where the first level of "catchball" takes place. The focus now is to understand how to achieve the goals defined in the previous steps, developing the set of projects necessary to achieve the annual targets and therefore priority for the entire company. Each project will have its own evaluation metrics and the KPIs will be those of the project team or department that will execute the project. This is the phase in which organizational alignment takes place, in which each team and each department integrates with the company's strategic plan, with a full understanding of its own contribution.

Among the projects to be set up, there must never be a lack of projects for building the **strategic competencies** and **management systems** necessary to realize the objectives as already seen in Figure 13.3. Returning to the example of expansion abroad, we will be able to direct and initiate a

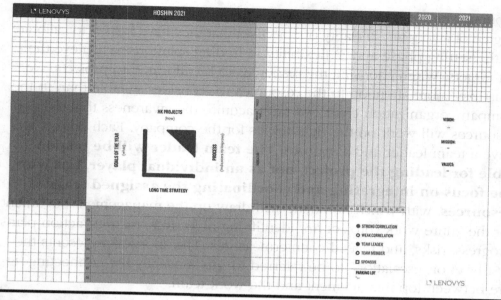

Figure 13.3 A3 X-Matrix format for strategic business planning summary.

specific project to build the necessary competencies: a new manager experienced in international development, specific sales networks, commercial ability to support the task.

The work of defining this phase can be considered completed when each annual objective is supported by one or more strategic projects, that is when we have closed the link between "what to achieve" and "how to do it."

Definition of Metrics and KPIs to Be Improved

This is the operational phase in which the contents of each strategic project are developed, using the A3 Management method (I will go into in more detail in a dedicated section). It is important for each team to share operational actions and specific metrics for each project to have an unambiguous frame of reference for subsequent monitoring. In the face of the huge availability of data and KPIs potentially available, it is important to keep a close eye on what we have chosen as a priority for the entire company.

Resource Allocation

Strategic projects both in the set-up and implementation phases are a fabulous weapon to get agile working and organizational methods off the ground. The work teams that are created must be small, with 1) a high degree of autonomy, 2) clarity of objectives that they have helped to calibrate in the setting phase and 3) flexibility in finding solutions to achieve the set targets. When choosing the resources to be allocated to projects, we must pay more attention to the resources we consider suitable than to the company organization chart, we must acquire the awareness that the right resources will work on the right things for the company. Each project will have a team leader and a sponsor. **The team leader will be responsible for leading the projec**t, **not as an individual player, but with the focus on integrating and coordinating the assigned team of resources**, with which he or she will draw up the management document for the entire work (project A3) identifying objectives, metrics, action plans, progress, risks, and related countermeasures. The sponsor represents the first level of "escalation" to be referred to. He is usually the connecting figure between top management and the work team.

Monthly Review Plan with Deviation Analysis and Countermeasures

Everything is set up. The corporate strategy has been generated and operationally translated into priority projects. We have defined what to measure and who will do what. The entire company strategy is enclosed on a page, the one in Figure 13.3, which has the great advantage of summarizing all the decisions made in the steps seen so far. Everyone involved knows what the long-term and short-term objectives are, what the priority projects are, how they will be measured and who is doing what.

Now the plan must be realized. The last piece of the plan is the one that can nullify the whole work done so far. This is the phase in which the project teams check whether the objectives of each project are being achieved and whether there are any blocking elements along the way. The monthly review is the meeting point between project operational teams and company management, so that any problems will not turn into insurmountable obstacles if not addressed immediately. The review plan will include not only the monthly appointments with management but also all the other organizational habits necessary to make the whole strategic machine work: project operational meetings, one-to-one meetings between team leaders and sponsors, "quick reaction management" rules to deal with urgent problems without waiting for the monthly appointment, and other forms of corporate communication and alignment.

Annual Revision

At the end of each year, the entire cycle is thoroughly reviewed and elaborated for the following twelve months. In this annual event, the objectives of the year that is ending are reviewed, new objectives are defined according to the contextual changes that have taken place and, above all, improvements to the overall strategy planning and execution process are reflected upon.

Lean Lifestyle Story: NT Food

Two Sisters and a Vision: To Bring Well-Being inside and outside the Company

> "My way of being an entrepreneur has had a total change starting the path of adhesion to the Lean Lifestyle Company model. For the first time in 30 years, I raised my head again and was able to focus on what really makes a difference for me and the company, and started a process of growing the organization and the people who work with me."

Speaking is **Nicoletta Del Carlo**, at the helm since 1989 with her sister Giovanna of Nt Food, a Tuscan reality, part of Morato Group, that today has three production plants and over 130 employees. A reference point in the Italian industrial food production for gluten- and lactose-free products.

The start of the change coincided with the Envisioning and Strategy Deployment process in 2019.

> For the first time, we aligned ourselves on the path we want to take in the future, sharing the goals and responsibilities of each project. Before the strategy was concentrated on my intuitions and that of my sister Giovanna, the results were the fruit of our abilities, but also of luck or external factors. In other words, success came many times thanks to gut feelings, instinctive decisions, personal visions grounded in one-way processes. Little was pondered and so I am certain that much more could have been done or much could have been avoided if we had adopted a structured method of governance as we are doing today.

The process of Hoshin Kanri Management has made it possible to empower people, to bring them on board really, and at the same time the ownership has been able to focus on the periodic review of improvement projects: "We are calm the work flows smoothly because everyone has shared the objectives and have clear on what to do."

"Yes, now I can say that I have no problem leaving the company whenever I want and be sure that nothing catastrophic will happen!" admits **Giovanna Del Carlo**. "With my sister we have gained time to devote to research and strategy, and of course to our mental and physical well-being. In such a difficult year, 2020, with a flat market, we achieved all our goals, and this was an extraordinary result because it is the fruit of a method. In such complex and difficult conditions, the proverbial elasticity and versatility of SMEs can no longer suffice. Getting results, valuing people, and taking care of their well-being has become a priority and cannot be left to chance if we want to succeed in surfing the high waves in which we sail.

Lean Lifestyle Story: Labomar

A Company Made of People

2017 is the year of the first Hoshin Kanri of Labomar, a company specializing in research and development and production of food supplements, medical devices, food for special medical purposes and cosmetics. Since then, every year, the strategic plan is reviewed, and innovation is driven through a new set of strategic projects involving the management and cascading the entire organization. Over the years, strategic projects have been gradually reduced in number to improve focus and organizational routines have been strengthened, declining the principle of the "inverted pyramid" into permanent cross-functional committees (Decision Support System, Operations, Logical Management, Scientific Commercial, Strategic) which, through meetings at least monthly, are fully responsible and autonomous for the management and control of both the current business and the progress of strategic projects. Each committee has gradually equipped itself with the necessary standards for efficient and effective management. This governance structure is not to be considered fixed and immutable, but subject to natural evolution. An example of this is the recent strengthening of the HR function, which, on the arrival of a new HR Manager and Labomar's increasing interest in investing in sustainability, has started the construction of a new committee dealing precisely with People & Sustainability, the Well-being Committee. Comments the CEO, **Walter Bertin**:

> At first, I had a clear vision and point of arrival, but it was not immediately easy to get all the people in the company on board. Today, four years later, the change I wanted has happened: people are proactive, helpful, and empowered. The proof? When we started the process of listing on the Stock Exchange, investors, after looking at the numbers, came down to production, they saw and breathed the enthusiasm, the serenity, and the smile. They saw the value of the Lean Lifestyle. They saw in the gestures and words of my employees that Labomar is truly a company made up of people. A goal that we defined back in 2017, and with it a roadmap that led us to develop vision and foresight, a focus on value, agility and simplification, the development of human potential, and the physical and mental well-being of our people.

More well-being and more results, without compromise. So in the difficult year of 2020, Labomar kept the bar straight and, thanks to the work carried out in the Lean Lifestyle perspective and its "training" to ride change, recorded a robust increase in revenue (+26.3%).

Manage Projects and Strategic Initiatives with a Gas Thread: A3 Management

Faced with a project to be launched, an initiative to be undertaken, a complex problem to be solved in private life, a checklist, a personal diary or even more simply a blank sheet on which to take notes with a bit of lucidity may be sufficient to clarify one's ideas on objectives and results to be achieved. Afterwards, it will be necessary to monitor everything with a bit of discipline. In addition, we may choose to be supported by guidance and confrontation with a so-called "sparring partner": a coach, a relative, or a friend we trust. **This choice will force us to explicate, objectify, and better clarify all the elements of our cognitive process and will increase the chances of success of our initiatives and plans**. In the company this shrewdness is not enough. In both face-to-face and remote working conditions, it is necessary to have a system in place that guides people in adopting the criteria seen so far for the correct setting of projects, business initiatives and complex problems to be solved. To work smart, it is necessary to adopt enterprise smart management systems, which are at the same time an operational process guide, an element of collaboration and social alignment, a facilitating and enabling tool. In this way, such systems will have the ability to steer people's behavior in the desired direction. If we want to build in our company a culture of work by objectives and work that focuses more on results than on endless tasks, in which it is easy to get lost every day, we must guide the necessary behaviors accordingly. Toyota has been using and continuously refining their management system for several decades to consolidate their corporate culture. A key element of their management system, often misunderstood in the West, is the so-called A3 Thinking. This is a method that many readers have already encountered in my book Lean Innovation,[6] which has already been applied by thousands of people in hundreds of Italian companies. I prefer to refer to this method as *"A3 Management"* to focus attention on the management and managerial aspects of its use. The name A3 comes from the classic paper format of 420 mm for the long side and 297 mm for the short one. At first glance, it might appear as a simple standard format for managing a project or a strategic initiative in the company. John Shook, former Toyota manager and current Chairman of the Lean Enterprise Institute, states that it can take years to correctly master this business tool in all its implications.[7] Let us try to understand why and, above all, let us aim to speed up the acquisition of this mastery. The entire method is based on the definition of a summary document in which the information on each individual project carried out in the company is collected. And this document can be declined in different forms depending on the field of application: problem solving, strategic project, personal development plan, etc.

The document accompanies all life phases of any type of project: from birth to storage, passing through all intermediate management and monitoring steps. So far, all in all, nothing striking. Follow me and go deeper.

Methodological Guide

The document is structured in blocks with titles, texts and key questions that guide everyone in the company using the same format. In the example in Figure 13.4, you can see that the inserted blocks force the team to set the project considering all the criteria of correct definition of the target objectives and the expected results, reflecting on the meaning of the project—the reason to act- and exploring the search for obstacles and interdependencies with related countermeasures. The document has four sections. The first two

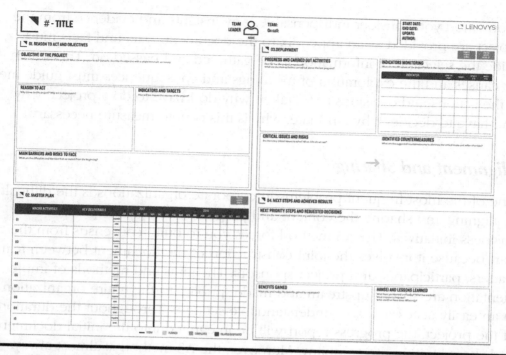

Figure 13.4   Example of A3 format for strategic project management.

are used in the design phase: objectives, reason for action, indicators and targets, prior analysis of obstacles and risks, planning. Sections three and four are used during the project: progress status, monitoring indicators, critical issues, risks and countermeasures, next steps and required decisions, benefits obtained, *hansei*, and lessons learnt. This structure guarantees concrete support to people and the active prevention of many "off tracks" caused by the wrong methodological approach to a project. You can simplify or enrich it, you can make different types depending on the complexity of the project, but there are three mistakes I suggest you do not make: a) do not use any standard document for project management, b) use too many different standards in your company, c) let everyone use their own standard.

Synthesis

Often in a company, when asked about the progress of a project, we could witness endless presentations with an infinite number of slides and information to search for the important ones. Or we may witness awkward explanations full of words told with little information and data to support them. The A3 method forces you to summarize, but above all to provide the key information you consider indispensable. Any insights and evidence from the project are provided as an appendix to the document. Data and facts must guide the selection of information to be gathered and passed on. The search for causes and the exploration of meanings and consequences must guide me in the actions and decisions to be taken why do I have to do a project, why do I have to decide, why am I late, why is this countermeasure necessary.

Alignment and Sharing

One of the most frequent problems in corporate organizations is the difficulty in aligning and sharing the content and progress of corporate projects and business initiatives. The A3 method forces you to build consensus from the start because it involves the joint construction of the document between team leaders, participants, and project sponsors. The aims and methods of implementation are shared upstream and made "public" to the entire organization in an easily accessible and understandable scheme. Throughout the duration of the project, the progress report will be integrated into the initial document and shared within the company. Moreover, the A3 method is almost always embedded in a broader digitalized system of strategic project governance, built to improve its use, management and sharing within the company.

Simplified Monitoring and Visual Management

One of the operating principles of a Lean company concerns the visual management of work environments: Visual Management. With great effort, several companies are striving to apply this principle in production departments. In theory. Unfortunately, billboards, signal lights and other pretty inventions are not enough to make a work environment visual. All it takes is to apply a simple rule: within a maximum of five seconds, anyone must be able to understand what is happening anywhere in the company. In production this will mean understanding in less than 5 seconds what has been produced against the intended target and in the event of an anomaly, immediately understanding why and who is doing what. Simple. Visual information and management equal vision. Companies often stop at visual information, but forget the most important piece, the transparency of visual management: Management. To build this second pillar, we need processes for analyzing information, assigning precise responsibilities, timely counter-measures, deep involvement of those directly affected and those with the levers to act are required. The same rules apply to the management of the "intellectual" activities of a company. It is often difficult to understand, in five seconds, where the various projects are in progress. Sometimes there is a lack of information, sometimes there is information, but it is not readily available, sometimes it is interpreted differently depending on who is processing the information. For Tom a project can be green, for Harry it can be red. The A3 method provides a simple tool for monitoring the individual project by means of traffic light indicators, the color of which does not depend on the subjective assessment but is linked to the elaboration of a simple checklist with the aim of making the information as objective as possible. Based on the traffic light assessment that has emerged, the definition of countermeasures and decisions to be taken becomes mandatory.

Coaching and People Development

The devil is in the details. The way in which a person processes the various elements of A3 will reveal the way they deal with problems and the way they think about their solutions. If used correctly, this tool provides a great opportunity to support people on their learning and growth path. This could prove to be the most delicate and at the same time the most powerful point of the method because it involves the evolution of the manager's role within the company. If the manager's objective is to make people grow by developing their potential, it will no longer be sufficient to "check" whether objectives have been achieved and tasks performed. Each moment of control must evolve into a moment of supervision in which, in addition to the results, the personal and organizational processes followed are evaluated; that is, in addition to what was been done and achieved, the "how it was done and why" must be evaluated. The A3 method facilitates this coaching process of the manager toward the employee, due to its structure and since the employee, together with his or her team, is obliged to pre-fill in the key fields prior to any time of checking on the progress of projects managed with an A3. The manager will have to carefully listen and read what has been prepared and ask focused questions that:

■ aim to raise awareness and responsibility on the part of the co-workers;
■ oriented toward defining new ways forward and new countermeasures to the problems encountered.

At the same time the manager will have to remove any obstacles, make appropriate decisions, request new actions, and project reviews. All this is realized when the manager accepts the dual challenge of developing his people and at the same time advancing projects and solving problems.

Corporate Culture

As already repeated in several parts of this book, the culture of an organization is the result of the behaviors of its people. Behaviors reveal skills, competencies, beliefs, and values. The choices that are taken daily within a company reveal its skills, competencies, beliefs, and values. The choices that are taken daily within a company show the underlying culture. Just like a living organism, the corporate organization can evolve over time. There is no human behavior that cannot be changed over time. Unless one is faced with serious clinical pathologies or profound value discrepancies. The task of every entrepreneur is to create and shape the desired type of culture in the company. The more homogeneous this culture is, the more synergies will be created preparatory to achieving great team performance. A3 Management offers the great opportunity to create cultural uniformity around a simple and systematic method: orientation to synthesis, objectives, structured problem solving, facts and data rather than words, sharing and teamwork rather than individualism. **When this culture pervades an organization, the good A3 will only be the superficial tip of the iceberg** (Figure 13.5). Under the surface of the water there will be people who, when faced with a problem, will stop, and systematically think about the consequences for the company, root causes, countermeasures, and action plans, rather than rushing headlong into the first solutions that come to mind. And above all, there will be people who will never play their games alone. The same people who, when faced with a strategic project to be launched, think first about the why, then the what, and finally the how, following a logic that can make the difference between getting results or not.

Lean Lifestyle Story: Akhet

Values Becoming Actions

> I first regained possession of my values and my vision, I stopped and really listened to myself, and then gradually clarified the many doubts I had as to whether it was appropriate to grow Akhet, the

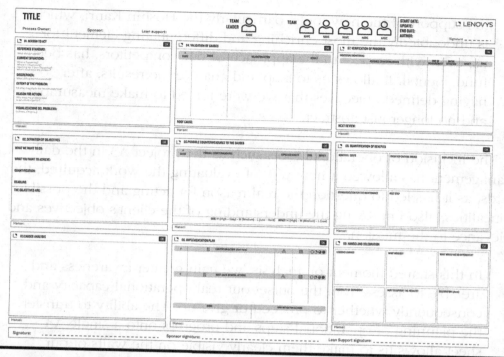

Figure 13.5   Example of A3 format for problem solving.

company that I had founded together with other archaeologists, almost twenty years ago, without any specific business training, to enable us to transform what for all of us was a passion into a job.

Going back over the Lean Lifestyle journey for **Claudia De Davide** means starting from the deep reasons that pushed her to find the time to delve into the methodology, first on an individual level and only later as an approach to business management. "For me, it was important to understand how growth does not necessarily lead to moving away from one's own values in order to adopt pre-established models, that there are different ways of doing business and that we can shape them to make our ideas and visions concrete."

After attending the Master in Lean Lifestyle Leadership, Claudia started the envisioning process, shared with the core team.

Although it was not a linear path for us and the results were probably not what we had initially expected, the process had the merit of helping us to clarify everyone's objectives, allowing us to realign our company project and project it over the long term. And in this, the support of the new tools, particularly the Hoshin Kanri, which is undoubtedly the most enriching and differentiating element for us, even with respect to the approach of our competitors, has been fundamental. It allows us to stop and analyze the results, after having defined objectives that we were forced to make measurable and no longer just abstract.

The inclusion of operational tools such as the project A3 in the daily management has allowed a new way of evaluating the work acquired, the yards, as it has led to question the real reason for acting and the possible criticalities, also improving the understanding of the client's objectives and relevance to the mission.

In this shared moment of analysis, we gain greater awareness and are able to assess from the outset our real operational capacity and consequently whether to proceed or give up. The ability to transfer the planning of operational steps and progress onto a single A3 sheet allows us to make them clearly visible on the walls of our office, facilitating governance and improving the sharing of results and team alignment.

Lean Lifestyle Story: Marcegaglia

A New Way of Managing the Company to Get Results

The Marcegaglia plant in Forlimpopoli is the great pipe factory that the group's founder called "the world university of stainless-steel pipe": 49 production lines and more than 400 "professors," to remain in the metaphor. The long experience of the organization and the high technical skills of the workers have led to excellent production results even during the crises of recent years and to an undisputed financial solidity that has followed a continuous growth of the plant. These records have, however, at the same time been an obstacle to the evolution of certain aspects of the plant. **Mirco Gugnoni**, responsible for the improvement, explains:

When we decided to embark on the path of the first Strategy Deployment at the end of 2018, the starting situation saw the greatest criticalities and weaknesses in the organization for silos, a portion of demotivated workers, high values of absenteeism compared to the group and the territory, low polyvalence, indicators of effective plants with wide margins for improvement, some obsolete machines, criticalities in the inventories of raw material and finished product. On the other hand, there were also strengths of which we were however aware: the leading position in the pipe market, the great production flexibility, the high skills of individuals and the professional ethics of the plant in reaching budgets.

The first activity of the strategic plan was the envisioning, with the involvement of the plant's key management figures and related functions at the Gazoldo degli Ippoliti headquarters.

"The 'fancy' exercise, which may seem simple, was not trivial for those who are used to reason with daily indicators, such as the tons produced," admits Gugnoni.

Therefore, the preparatory phase was a question-and-answer session for individuals, on examples of work ethics, professional qualities, major work successes and defeats, and what we liked doing in the company most. Precisely related to the last point was an exemplary moment to describe the working climate among some of our departments. To the question "What can't you do without when you come to work?", one of the participants answered: "Taking revenge on colleagues who do me wrong!" In the end, it was here that the idea of the direction in which we should go began to germinate better relations between services and production, more empowered people, a more layered organizational chart to avoid bottlenecks in decision-making, reduced warehousing, reduced breakdowns and increased speed.

At this point, the Vision at three years has become clear: "To become a team of motivated professionals to provide customers with the best steel solutions developed in a model company for its territory."

The point of union between "who was" the company at that time and "who would become" at the end of the journey, was the Marcegaglia Strategic Plan 2019–2021, formalized in an X-Matrix. Gugnoni tells again:

The projects, which over the three years followed one another, would hardly have been able to reach the budgeted targets if there had not been adequate governance and an escalation system upstream. In fact, each project has a team leader and a team that meets weekly to define and monitor project progress activities. Every fortnight the team leader meets the PMO of reference who, with a maieutic approach, controls the progress of the project and provides support to the team leader in case of need, for example on obstacles that the team leader with his hierarchical position cannot remove.

Every month, the PMO meets the management and team Leaders of the "red" projects, that is those with critical issues on results or progress of activities that not even the PMO was able to remove and where Management intervention is required. The last level of governance is the meeting every three months with the steering committee composed of the highest figures of the central organization of Marcegaglia Specialties (Ownership, Commercial Management, Human Resources Management, Quality Management and Plant Management), where each team leader presents their project in a meeting of a mainly celebratory nature.

This is the appointment that can give the team leader great motivation, put him in front of the rejection of the project or show him the need to back up to the starting point, but in any case, it is an opportunity to collect advice and start with energy. Gugnoni recalls: "Some episodes have become legendary, such as when a Team Leader turned to the owner and said 'Doctor, I understand you' in an amusing role reversal."

The organization of the strategic plan has brought great improvements, especially in terms of structure and method, so much so that some tools—such as, for example, the A3—have been disseminated to all the plant's activities.

Results?

Not all the improvement projects achieved their objectives, but the vast majority of them led to extremely positive results: an 8% increase in the speed of the plant's largest production department; a 30% average reduction in the set-up times of two departments; a drastic reduction in obsolete and slow-moving materials in the warehouses; an improvement in the plant's microclimate and in the ergonomic situation of the most critical workstations; and a 2 percentage point reduction in absenteeism.

Lean Lifestyle Story: Streparava

Hoshin Kanri: The "Navigator" in the Company

In these times of high variability, making medium- to long-term strategy appears to be an effort with uncertain outcomes. This is well known at Streparava, a leading company in the automotive components industry, which, to face the dynamics of an extremely changeable sector with lucidity, has adopted the Lean Lifestyle Company model for some years now also at the level of corporate governance and strategy. Here is how **Paolo Streparava**, CEO of the company, experienced the use of the strategic planning processes and tools described in this chapter.

> When you are in the car and you want to go somewhere, you use your navigator; so for any reason, if there is a blocked road or there is traffic, you rely on this tool to recalculate the route to your destination. We chose to have our navigator identify the goals and how to measure their achievement. Through the Hoshin Kanri we charted the course and even though our route was hampered by the arrival of the Covid-19, we never lost sight of our final destination on our radar and adapted accordingly.

Despite the stormy sea, although it is difficult to predict the future, the focus at Streparava was on getting the company and people toward the goal. This is the first task of a so-called Lean Lifestyle Company that wants to bring maximum value with minimum waste of time and energy and with maximum utilization of its human and technological potential. Vision and foresight are the first ingredient that a company today must use more.

> Our journey started in 2017 with the envisioning process that allowed our employees to feel in their gut what they wanted to make happen in the company where they worked. This was a very important element for Streparava, which had already tried its hand at vision and mission definition in the past. We had also done it with the help of competent people, but this vision and mission seemed more a declaration of intent than a deep feeling. So, the exercise we did with the help of Luciano and his entire team was to involve people in the future design of Streparava. I believe this is a truly indispensable experience for a company. If people feel in

their gut the company, its future, their role as protagonists in the realization of the final goals, then we will be sure that they will all row in the same direction.

Being a Lean Lifestyle Company in Covid-19 meant in Streparava never lose the compass, focus and well-being:

> First of all, we have secured our people, a key priority for us and consistent with our values. Then we took our shared strategic plan, the HK, and we updated it with the energy that only those who feel the company's projects and goals on their skin can demonstrate.
>   Is this enough? No, we need constant alignment.
>   It is not enough to make a strategic plan, it is not enough to generate the company's mission. If this is not followed up every month, every week, through a monitoring system involving people, it becomes difficult to keep the whole team cohesive and compact.

Also in this the Hoshin Kanri is a winning tool, Applied in Streparava:

> It's a way of working that costs effort at first but pays off in the end. My team and I now know at all times how projects are progressing and when the Covid emergency had to be dealt with, it was easy to calibrate them for a new scenario. Being a Lean Lifestyle Company means taking off the fireman's helmet: how many companies and managers in Italy excel in this habit of working, in good times and bad, with a sense of urgency and "rescue"?

Digitalizing the Entire Strategic Process: NEYM

The numerous experiences accumulated in the implementations of what we have seen in this chapter have led us to consolidate the processes and tools analyzed in a digital platform. The idea that took shape was to create a lean digital solution to support any company wishing to equip itself with its own strategic planning system, in line with the principles of a Lean Lifestyle Company, driving all the execution phases of the strategy. NEYM (LeaN LifestylE CompanY SysteM) is a lightweight system, fully adaptable to one's needs and incorporating the key elements of Hoshin Kanri, A3 Management, agile governance of the entire strategic plan. A true digital navigation system to quickly build, manage, execute, and change the company's strategy along with the projects and activities needed to achieve it.

If you want to know and explore the details of the digital platform for strategic business planning and execution scan this QR code and visit the NEYM website.

Notes

1. Simon Sinek, *Start with Why. How Great Leaders Inspire Everyone to Act*. Penguin Books, 2009.
2. A. G. Lafley and R. L. Martin, *Playing to Win: How Strategy Really Works*. HBR Press, 2013.
3. Salim Ismail, *Exponential Organizations: Why New Organizations Are Ten Times Better, Faster, and Cheaper Than Yours*. Diversion Books, 2014.
4. J. M. Kouzes and B. Z. Posner, *The Leadership Challenge: How to Make Extraordinary Things Happen in Organizations*. Jossey-Bass Inc Pub, 2012.
5. www.brightline.org/resources/eiu-report/
    Special Report—Closing the Gap: Designing and Delivering a Strategy That Works. The Economist Intelligence Unit, 2017.
6. Luciano Attolico, *Lean Development and Innovation. Hitting the Market with the Right Products at the Right Time*. Routledge, 2019 (pp. 170–175).
7. John Shook, *Managing to Learn: Using the A3 Management Process*. Lean Enterprise Institute, 2008.

Chapter 14

Becoming a Lean Lifestyle Company

You have reached the last stage of your journey through the pages of this book. If you have come this far, it means that your interest in the topics covered has been high. Now it is up to you to decide what to do with all that you have collected so far: concepts, methods, tools, reflections, examples, testimonials. I hope I have contributed to new awareness, to have enriched your knowledge and have triggered your personal reflections.

And once you have finished this last chapter, you will have some options. Go back to work exactly as before, perhaps thinking that, if the company does not change then there is no point in changing the way you work. Postpone the transition to action until you feel like it or think there will be more favorable conditions. Experiment on a personal level some of the techniques and methods that have impressed you most or that you consider most applicable. Start a structured self-improvement project with the aim of bringing about some tangible changes in your personal and professional life. Start structured Lean Lifestyle projects in your company to work better and achieve the goal of increasing performance and well-being.

Whatever action you decide to take, I want to give you some good news and some bad news. I'll start with the "bad." You will not be able to keep the work and personal dimensions separate. You work most of your active time, so it will be difficult to think about improving anything in your life while leaving the professional sphere untouched.

The energies at play in your work sphere are so impactful and pervasive that they inevitably end up permeating your personal sphere. If there are

DOI: 10.4324/9781003474852-14

working habits that are not conducive to high performance and well-being, it is certain that they will also generate negative consequences in your personal life. And vice versa.

The first good news is that, on an individual level, every small change you make at work will inevitably have major consequences for the rest of your life. Experiencing deeper relationships, finding meaning and purpose at work, reducing stress levels, getting used to a high-focus, high-impact way of working, saving one or two or three hours of work each day, could have a real impact on your life that you may not even be able to imagine.

But to achieve these results, there is no program or business organization that holds: you need your convictions, your commitment, and your concrete actions.

On the other hand, at the corporate level, I think it is now clear that there is an indissoluble link between what the company decides to do at the organizational level and the profound impact on people's performance and well-being. Entrepreneurs and managers have a huge responsibility in this regard and often, unfortunately, remain tied to business models that are no longer functional to what a rapidly changing world requires.

The other good news is that the link between people's well-being and company performance has been proven by multiple scientific sources and numerous empirical testimonies. Investing time and energy in this, therefore, has a tangible return for people and companies.

Corporate and Individual Systems in a Dynamic Context

The experience I have gained over the years and which I have tried to transfer to you with this book leads me to be increasingly convinced that people and companies must be linked together in a synergetic and complementary way. Only by moving together can we build a complex system capable of adapting readily and effectively to a constantly and rapidly changing environment.

A widespread paradigm has caused a lot of damage: that of seeing companies as a single system, made up of organization, maps, procedures and flowcharts, machines, plants and all the other tangible corporate elements. There are two systems in companies that coexist: the corporate one, that is everything that we can objectivize and codify as real, measurable, and observable company asset, and a set of individual systems that cannot be objectivized or codified exactly, but which can dramatically increase the value of any company.

Each person's behavior is never completely predictable and replicable. I cannot say that I really know how objectively a person works, what he does every day, what micro-habits guide his choices. If I want to achieve measurable and tangible improvements at the individual system level, I must use different methods and practices than what I can do at the company system level. And it is not possible to profoundly improve a company if we do not simultaneously intervene in elements of both the company system and the individual system (Figure 14.1).

I think that obsolete change management practices that are geared toward "convincing" people to "overcome their resistance to change," as in the case of adopting a new IT management system or new business processes and organizations, typical elements of a company system, should be abandoned once and for all. People cannot be seen as a burden to be pulled

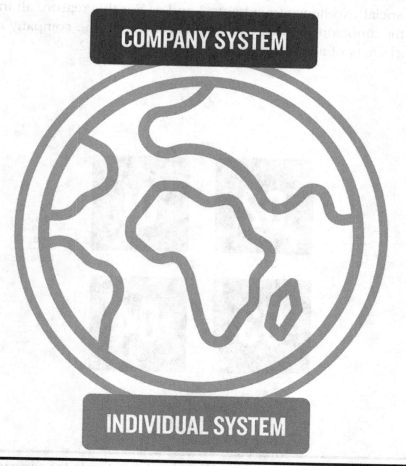

Figure 14.1 The two interacting systems in the company.

by force into a much higher and distinct corporate change design. The dual interactions between business systems and individual systems will profoundly influence the fate of the company.

You need to add another element of the puzzle.

Excellent results come from companies, made by excellent people and processes, individual and corporate systems. The results obtained derive from two intertwined forms of excellence: technical and social.

Often companies focus only on the most observable form of technical excellence (quadrant 1, Figure 14.2), which goes well with the definition and development of business systems, that is, with everything that can be measured in a tangible way, but which does not ensure sustainability, well-being, consistency of performance, and sometimes does not ensure the expected results.

For this purpose, it is necessary to consider both forms of technical excellence at the individual level (quadrant 2), as well as forms of individual and corporate social excellence (quadrants 3 and 4) For this reason, all initiatives that have the ambition to bring the Lean Lifestyle into the company refer to four distinct areas of action, as represented in Figure 14.2.

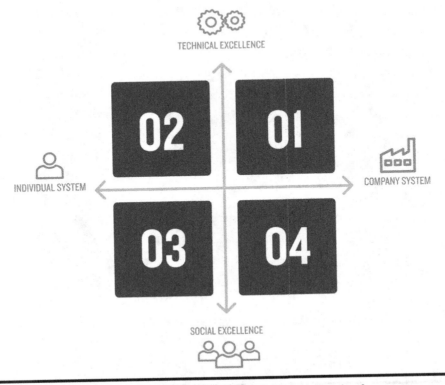

Figure 14.2 The action areas of a Lean Lifestyle intervention in the company.

Dial 1. Company System/Technical Excellence

Quadrant 1 will include implementation projects of the purely technical nature (a new management system, a new layout, a technological upgrade, a new plant). We are talking about all those projects whose results strictly depend on elements of a technical, measurable, and objective nature. A structured product or process cost reduction method, for example, intervenes with tangible corporate factors and needs technical excellence to enable significant results to be achieved.

Dial 2. Individual System/Technical Excellence

Quadrant 2 includes projects without which it is difficult to support those of the first quadrant.

And I experienced this it myself a couple of years ago. Paolo, my dear client, and friend had just bought a fantastic blue Ferrari. I can assure you that if you are not used to or accustomed to driving such a car, you cannot understand the difference between that Ferrari and any comfortable sedan. When Paolo asked me if I wanted to take a ride in the new car, I gladly accepted with childish enthusiasm, unaware of what I would experience. OK, I start the engine, drive off and immediately a wave of fear mixed with panic envelops me. Acceleration and driving mode were devastating for me. On the road I was overcome by the urge to push on the accelerator, but terrified by the fear of crashing into the corners. The moral of the story, I realized firsthand that to enjoy the pleasure of driving such a car, you must first learn to drive it well. Very well. And you must feel fit to drive it in all conditions.

The same thing happens in the company.

Any evolution of a part of the business system needs its human counterpart to drive solutions toward the expected results. In quadrant 2 are projects that lead individual systems to adopt the forms of technical excellence necessary to make the whole work harmonious. In quadrant 1 there is the Ferrari, in quadrant 2 there is the Ferrari sports driving course and the evolution of the person's identity from driver and car driver to sports car driver.

Quadrants 3 and 4 contain all projects that touch on the relational sphere. As I said earlier, it is the interactions within the company that will determine its success or not. Quality, effectiveness, and synergies of interactions are fundamental to achieving what I call social excellence.

Dial 3. Individual System/Social Excellence

Quadrant three contains the projects of individuals designed to evolve their personal contribution to the social excellence of the entire company. Delegation, communication, work–life balance and harmony, the development of individual social skills, self-efficacy, self-organization, are all examples of individual developments on which the company can only intervene at the level of promotion and incentive, but on which the real implementation power lies solely in the hands of the individual.

Dial 4. Company System/Social Excellence

Finally, in quadrant 4, you will find examples of corporate, and therefore well-coded, objectively observable, and measurable projects that aim to move people and companies toward social excellence. These include strategic alignment projects, projects to promote dietary and physical well-being, projects to define motivating and engaging Mission, Vision and Values, smart layout projects, and projects to implement new organizational habits, such as the ATRED. In Figure 14.3 you can find some examples for each of the four quadrants.

Lean Lifestyle Story: Labomar

Toward a Lean Lifestyle Company

Labomar, a company founded in Istrana in 1998 by a pharmacist, **Walter Bertin**, with the dream of creating natural products for people's well-being,

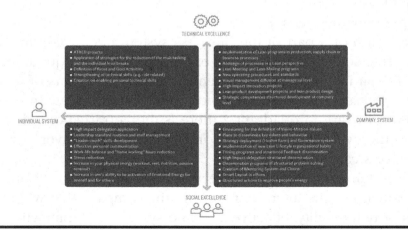

Figure 14.3   Examples of Lean Lifestyle projects in the company.

is today a reality specialized in the research and development and production of food supplements, medical devices, foods for special medical purposes and cosmetics.

The meeting with the Lean Lifestyle takes place in 2017.

"When I first heard Luciano Attolico talking about Lean Lifestyle at a conference in Vicenza, I said to myself: this is stealing my words!" says the CEO of Labomar. "I was fascinated by the will and the ability, in which I believed, to combine company performance and people's well-being without compromise."

Thus began a process of change that went far beyond the normal scope of Lean tools. A program designed to increase both the company's performance and the energy of its people at the same time, with an extensive series of interventions that have contributed to the company's ongoing rapid growth in terms of production, turnover and geographical expansion.

A new way of working, new corporate governance tools, combined with the foresight of its founder and director and the commitment of more than 200 employees, are bearing fruit, and so the company has broadened its horizons, so much so that in 2019 it acquired an industry company in Canada and on 5 October 2020 it completed its listing on the Stock Exchange. All this without losing its "true north": improving the well-being of people inside and outside the company.

Labomar's journey in its Lean Lifestyle started with the Hoshin Kanri and the establishment of cross-functional committees and continued with several key steps. Let's see them.

Roles, Responsibilities, and Growth of People

In several areas of the company, roles and responsibilities were revisited to support a culture of delegation and the continuous growth of employees. In production, for example, the traditional activity of coordinating the department manager was modified to refocus it from an excess of scheduling and documentation activities toward a prevalence of higher value-added activities, such as process innovation and the growth of employees. In this case, the approach was ATRED, to understand the activities of the day, sizing them by average time spent, and defining directly with them which of these activities were gold for the Head of the Department, and which were instead delegable to more operational roles. This work was then extended to the entire organization, revised based on the strong impetus we wanted to give to the processes of innovation and improvement in the department. This led to the creation of the figure of the Production Team Leader, a key role within the organization, who combines strong technical knowledge and experience of machines with a decisive focus on improving performance in production.

Lean Agent Labomar

As part of continuous improvement in the company and innovation of internal processes, Labomar has undertaken the identification and activation of 11 cross-functional figures called Lean Agents. Lean Agents, also called Change Agents, have a clear objective in the company: to find waste process and apply themselves to their removal, always chasing maximum value for the customer (internal and external). To pursue this, the people identified were engaged with training to pursue personal excellence as their initial goal, believing that if we become capable of change for the better on an individual level, we will be more easily able to guide others to do the same. In fact, when Lean Agents acted on other projects, beyond their own personal scope, their example, enthusiasm and support was appreciated by all.

High-Impact Training

In Labomar, a training plan was implemented starting with the first- and second-line managers, with progressive extension toward the other company operational figures, to develop skills in the thematic areas covered by the project. The effectiveness of each training session was measured through evaluations and tests that participants took at the entrance and at the end of each module to appreciate the real improvements achieved. Classroom training was followed by follow-up activities and applications in pilot projects.

"Only in 2020 we had 1446 hours of Lean Lifestyle training and 114 hours of testing, with 111 people involved, while in 2019 there were 63."—says **Gabriele Schiavina**, Chief Transformation Office at Labomar. Training is not only in the classrooms, but the company is also investing heavily in increasing people's energy.

"We opened a company gym," Walter Bertin continues, "available to employees and hired personal trainers who observe people at work to help them maintain correct posture. In addition, we also do training on the topic of proper nutrition and offer company appointments dedicated to sports throughout the year."

Revision of the Product Development Process

The new product development process was reviewed in search of all those actions necessary to reduce response times to customer requests. Often in this world, a simple delay in response, quotations, offers, feasibility studies, samples, and prototypes, can mean losing a possible new order. The process mapping work in this case was combined with the ATRED analysis of the individuals involved, and the result brought to light numerous opportunities to review processes, tools, and organization, also taking advantage of them to take steps toward digitizing certain working tools. As, for example, was the case with the "project map" that documents that ferries customer requirements from sales to R&D and follows all the stages of the newly defined process, from the formulation to the final product development. In this case, a standard digital document was created which, by communicating automatically and continuously with a dashboard, makes it possible to monitor the progress of each new project throughout the entire development process. At the same time, the conditions were created to assess in real time the effectiveness and efficiency of the various phases in which the value stream of new product development has been revised: Concept, Development, and Industrialization. The new tools enabled the newly defined processes, and in particular:

■ They have facilitated people in routines, i.e. in the creation of standard (digital) documents for faster communication between departments and working groups and faster compilation of the same.
■ They made it possible to monitor the results at any time, to identify any critical issues and immediately find the most appropriate countermeasure to the problem.

To ensure the agile operation of the new flow and to guarantee accountability and downward delegation, the existing roles were reviewed, new ones were defined, and new management figures were selected from among the highest-potential operational people, giving them the prospect of significant growth. This made it possible to make the organization functional to the new process and orient it toward results, which were not lacking:

- reduction of first response lead time (concept) by 50%.
- reduction of customer response lead time by 20%.
- potential increase in conversion rate of 100%

The Supply Chain Process Review

Given the excellent results obtained with the new product development process, the company decided to implement the same methodological model to the supply chain. Here, too, process mapping was combined with ATRED analysis of the activities and workload of individual people.

Depending on what emerged from the information collected in R&D, they decided to focus the working group's attention on optimization, standardization, and digitization of documents.

Improvement of the Production Process

The work site covered the analysis of the current production process, the identification of the main areas of intervention and the application of countermeasures through the work sites led by the new production team leaders. Some examples of improvement actions: quick setup to reduce machine idle hours during the various types of setups (batch change/format change/ product change); quick reaction meeting to reduce rejects and rework; daily meeting to establish daily priorities and define countermeasures to current criticalities; visual management to provide real-time visibility of deviations between production carried out and expected, reporting the causes and related countermeasures.

"The investors," Walter Bertin concludes, "did not believe in chatter, but in what the team demonstrated during the production visits: they found people who cared about their machines, enthusiasm, smiles and serenity. They experienced the Lean Lifestyle in the field, and this was rewarded. They understood that Labomar is a company made of people."

Lean Lifestyle Story: Streparava

A Family Since 1951. Ready for the Challenges of the Future

Streparava is a reference player in the automotive sector producing high-precision Powertrain and Chassis systems, mainly dedicated to the Premium Car, Light Commercial Vehicle, Bus & Truck and Motorbike markets. With a consolidated turnover of more than 230 million Euro, 930 employees, and 7 plants in various countries around the world, the company supplies some of the world's leading car manufacturers such as CNHi, Ferrari, Ducati, Mercedes AMG, and Porsche.

The Starting Point

Paolo Streparava, CEO of the group and third generation of the family at the helm of the company, is a man who hardly stops at an obstacle and who loves to fight for what he considers the "right thing to do." Here is how he describes his decision to transform the company into a Lean Lifestyle Company.

> I have always felt a strong desire to bring changes to the company. In me there was the desire to understand how to contaminate a way of running the company that was already successful anyway; how to do even better and above all how to make our capacity for inventiveness, adaptability, and energy systematic. Since 2015 we had started a continuous improvement program in production based on the WCM (World Class Manufacturing) methodology, and we had made several attempts to introduce new methodologies, such as the Theory of Constraints. But I was still not fully satisfied. I was always missing something and gradually I convinced myself that a Lean Turnaround was the way to create the gear shift I was looking for. The decision-making process with my team at the start of this journey, however, was a long process: after a year of sharing and trying to convince, there was still no end in sight. In the Lean Lifestyle, in Lenovys and in Luciano, I found the ideal methodology and the ideal partner who could marry deeply with the founding value of our company and our daily operations, the focus on people, and who could help us strengthen it.

Paolo tells how he decided to start this journey.

> To achieve this transformation, the "raw material" I thought I would act
> on was my team. 14 Mondays: the front line of the company studied
> the Lean Lifestyle eight hours a day, me first, for 14 Mondays. All our
> employees saw that we spent entire days in the meeting room study-
> ing. It was hard because we challenged each other.

An intense training that was immediately turned into action through individual projects and company-wide pilot projects, which began to spread the
Lean Lifestyle principles and immediately brought the first successes.

Vision and Execution

The next step of the Streparava journey was a real turning point. Through
the envisioning process, Paolo and his first and second lines defined vision,
mission, and values, and then translated them into a precise corporate strategy, as already told in Chapter 13.

Hoshin Kanri's process ensured that the defined strategies were pursued
month after month, thanks to a system of continuous involvement and alignment that allowed the team to preside, adjust the focus and keep the team
together.

The learning curve lasted a few years, but now, in the fourth year of
application, "in my opinion everything worked excellently," Paolo explains.
"We have learnt how to use the tools, but above all to understand how they
can help us shape our strategies."

Two were the most important hubs in the learning process: digitization
and the committee system. The complete digitization of the process made
everything simpler and smoother, allowing people to focus more and more
on the process alone. The introduction of the committee system, through a
downward delegation of all Hoshin Kanri project management, has made
the governance of the process far more effective.

> The committees were an organizational change that brought huge
> results: it simplified life and brought an acceleration of the abil-
> ity to achieve high business performance. We created four of
> them, focused on the key areas for our company: Operations,
> Commercial, Innovation and "SBS, People & Energy," focused
> on people and their energy. I have delegated to them the

responsibility of presiding over the Hoshin Kanri ("HK" as we call them) projects in their area. They oversee overseeing and intervening if a project is slowing down, has problems, needs "energy," or if it is finished and can be archived, having my "CEO Committee" as an escalation point if needed. I have entrusted them with a great deal of responsibility, and this could bring with it the fear of losing control of the company, but this is not the case because people know the objective, the strategy and how to carry them out.

It is a value-oriented organizational model with small teams entrusted with full responsibility for the process. With this change, Streparava has certainly taken a step in the direction of becoming a "value-oriented organization" and in overcoming the culture of functional silos that often generates unnecessary conflicts and does not guarantee maximum value in the shortest of times.

The Power of Habits

To help people express their best and realize their full potential, it is not enough just to act on an individual level: Streparava has elevated the Lean Lifestyle to a pillar of corporate strategy and has gradually introduced a series of new habits and systems into the company in a structural manner to help it realize this in practice.

> Habits are fundamental. If we do not create a positive result, it is not said that in the future this result will be repeated. One of the new habits introduced with the Lean Lifestyle that I am particularly fond of is the regular One-to-Ones, a time when our managers at all levels meet the people they work with. This has enormous power because people often tend to shut down because they don't feel listened to, but when this happens, after perhaps an initial moment of venting their anger or fears, they slowly start to contribute because they feel respected and valued. The One-to-One habit was then joined by another key habit: the periodic use of the ATRED, which has become a standard for all, including me, to identify Gold Activities and try to understand what improvements and levels of delegation need to be activated.

We have also trained ourselves in the Lean Meeting habits, which apparently seems like a small thing, but if I must judge by the effort, we put into maintaining the rules we have given ourselves I have to say that it is probably a bigger challenge than it seems. Another organizational habit that I find very powerful has been the use of Project A3 as a standard method for setting up all the projects carried out in the company, which is fundamental for getting people on target.

An interesting example of designing a new organizational habit was the introduction of the sacred time, as already seen in the testimony of Raffaella Bianchi in Chapter 5. In addition to planning in detail this time, a standard content structure for the five weekly slots was also defined:

- A personal Improvement slot dedicated to ATRED and personal improvement actions or Lean Lifestyle training and study on the Lenovys Lean Lifestyle® Academy platform, which is available to all employees.
- Three Gold Activity or Top Priority Slots
- A planning and *hansei* slot on Fridays

The introduction of this new habit was also monitored through a Habit Tracking smartphone application (Figure 14.4) specially designed to allow not only monitoring of the habit by individuals, but also cross-visibility of colleagues' results and the possibility of creating company-wide summary dashboards to get an overall picture of the evolution of the situation. Tracking lasted for about 12 weeks until the adoption rate stabilized at a satisfactory level.

Another habit that took root immediately and very well in the company were the Function Alignment Meetings.

Each functional team meets weekly for about an hour in front of a visual scoreboard set up on the logic of pull planning, which each function has designed according to its specific needs. Activities and priorities are shared, planning and workload decisions are made, important information is shared. This created greater team cohesion, better communication and sharing of what was happening in the company, alignment of priorities and early identification of problems.

From the point of view of habit design also in this case it was decided to synchronize the slots at the company level by scheduling them all on

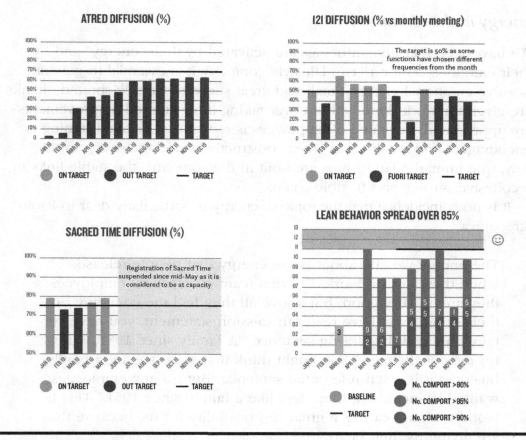

Figure 14.4   Dashboard tracking rates spread of new Lean Lifestyle habits.

Monday mornings. In this way, all functions make the alignment meeting at the same time slot and in fact there are no cross-functional meetings in that time slot: this eliminates the annoying effect of people being absent "for another urgent meeting" and in fact makes it a little easier to find people available during the rest of the week for cross-functional meetings.

In addition to these, Streparava has over the years built and spread many organizational habits based on Lean Lifestyle principles, such as Manager's Position Agreements, the annual calendar of all company routines, daily meetings in production areas, mail slotting, and many others.

All habits are planned, then disseminated and monitored in an organized way. The SBS, People & Energy Committee also monitors, among others, the KPIs linked to the introduction of the main new habits and intervenes to remove obstacles and correct the shot.

Energy in the Company

We have seen that human beings are activated by their "energy" and by their emotions, so for a Lean Lifestyle company it is essential to generate positive emotional energy. Small and great successes are celebrated, thanks are given for work done without ever taking it for granted, improvements are appreciated, the value of employees is recognized, active listening is encouraged, and feedback is given constructively and incisively. In this way, true human relationships are built in the company, the stable links in a cohesive, strong, and flexible chain.

It is no coincidence that the topic of energy is particularly dear to Paolo Streparava.

> The energy we talk about is the energy that people release when they have to work. I have a team of excellent employees: they are all very good, but above all they feel the company on their skin. If you have read our mission statement, you have certainly noticed the closing sentence, "A family since 1951, ready for new challenges." You might think that since we are a family business it is a self-referential sentence. But no, my employees wanted to include it: they "feel like a family since 1951." This is wonderful and carries a great responsibility for us, because they are giving us their heart.

The Streparava mission is:

> We provide high-impact solutions worldwide through the design, validation and production of driveline, chassis and powertrain components and systems.
>   Our people are committed daily to the safety, continuous improvement, reliability, and the well-being of all stakeholders today and tomorrow.
>   A family since 1951.
>   Ready for the challenges of the future.

However, energy is generated not only thanks to the setting of individual and group relationships, but also in the way the company is managed operationally and the levels of empowerment and autonomy that people are given (Figure 14.5).

Figure 14.5   The Streparava team: A Family since 1951. Ready for the challenges of the future.

"Listening to people, giving them trust, lots and lots of trust," Paolo points out. "The people you trust can bewitch you, and certainly the greatest satisfaction I have had over the years is the result of the trust I have given. I share with them the why, not the how. Sometimes I get angry if I see that in my collaborators there is not the same willingness to give delegation and trust. After all, if we have grown as a company, if they have grown as managers, it is because they have been given an enormous amount of trust that has allowed them to develop their qualities and put them at the service of the company. Leaving room for people certainly takes some risks, at least at the beginning; but everyone makes mistakes, entrepreneurs, and managers alike."

It is therefore no coincidence that Streparava has chosen "Physical, Mental, Emotional Well-being" as one of its three values and that it is doing a lot for people's energy. The other two values are "Team Spirit" and "Challenge."

This has led to numerous projects and initiatives: yoga courses, improvement of the company canteen, agreements with gyms and swimming pools, the football team, the cycling team, snowshoeing group, the "family factory," training focused on how to give feedback or how to increase emotional energy, just to name a few.

"It is important to transform all these values into projects, otherwise if it is not a project and if it is not measurable, it does not work," explains Paolo.

Leader Streparava

As a corollary of all the effort made to put all into practice all the principles of the Lean Lifestyle, Streparava has also crystallized what this means in practice for people, by defining and formalizing the key behaviors of the Streparava leader:

1. Focus on Gold Activity to achieve maximum results with minimum effort.
2. Plan one's own activities and those of others, to achieve the desired results in the set time.
3. Respect company standards and contribute to their continuous improvement.
4. Methodically address problems by basing decisions on factual data verified in the field.
5. Provide timely and constructive feedback to improve processes and people.
6. Collaborate actively with others to achieve goals.
7. Delegation, empowerment, and guidance according to the corporate values and principles of the Lean Lifestyle® Streparava.
8. Explore new avenues and is open to change to evolve the Streparava group.
9. Knows how to strike a successful balance between personal and work life, promoting physical, mental, and emotional well-being.
10. He teaches by example and generates positive energy for himself and others to grow.

The story in Streparava's Lean Lifestyle has taught us that the journey is more important than the destination: if along the way you can collect smiles, lessons learned from experience, mutual understanding, compassion, and respect, what you do along the way stays in everyone's heart and rewards more than anything else.

Lean Lifestyle Story: Lucchini RS

When You Break the Balance, Everything Is Restored. It Is Only a Matter of Time

The Lean Lifestyle project in Lucchini RS in 2020 did not go well. After about two years of activity, with mixed results, the word end seemed to have come due to dissatisfaction from various parts of the company. Something was not working, but we were struggling to figure out what.

The entire front line of management still had in their hearts the emotions of the envisioning process, where corporate vision and mission were written and linked to the company's long-term strategic goals. We had introduced the LMS—Lucchini RS Management System—as a single corporate platform to guide the path to corporate excellence, consisting of different pillars including Goal Alignment, considered the new model in corporate alignment.

We launched ABP (Agile Business Process) and high-impact innovation initiatives.

We have implemented many tools and achieved results in different business areas. Meetings and e-mail management were improved through the Lean Meeting and Lean Mailing programs.

Many managers and employees had made the Lean Lifestyle Leadership training and mentoring courses, with a focus on personal excellence.

But something was not going in the desired direction and there were several weak signals that, having been dormant and misinterpreted at the beginning of the project, were about to explode.

It is at times like these that you must stop and reflect to act differently. The company's top management team, including Managing Director Augusto Mensi, did not shy away from this deep reflection—*hansei*—to explore the causes of this looming "crisis" together with possible countermeasures, to relaunch the project and restore momentum and energy to all those involved, even in the face of the spreading Covid-19 pandemic.

Lucchini RS is an Italian family company with a long history behind it. It was founded in the immediate post-war period by Luigi Lucchini, the son of a blacksmith who worked iron in Casto, a small village in the Valle Sabbia, in the province of Brescia. Starting from his father's craft workshop, Luigi Lucchini has expanded production, especially in the field of rebar for reinforced concrete. Through investments and acquisitions, the small business grew to become Italy's second largest steel producer in the late 1990s. With an underlying philosophy: always reinvest profits back into the company. Throughout its history, the Lucchini group has always focused on higher value-added production, such as special and high-quality steels. The crisis in the steel industry at the beginning of the century led the family to disinvest from the mass production of rolled products and to focus on cast and forged products, coinciding with the Lovere Plant (acquired by the State in 1990) and its subsidiaries. This is how "Lucchini RS" was born, the meeting point of the Lucchini family history with the centuries-old tradition of a production site founded in 1856.

The sider mechanical group offers a variety of products and high-tech services on a global scale. The group's business activities are identified in two divisions:

- Production of high-end railway components;
- Production of custom-designed forgings and castings for various applications in the energy, oil, cement, automotive, naval, and heavy industrial sectors.

The entire production process from steel to finished product is in Lovere (Bergamo) (steel making, forging, foundry, mechanics, painting, and assemblies), but the group also includes companies in Sweden, South Africa, Belgium, England, Poland, Austria and China in addition to Italy, India and the United States. With a turnover of €424 million in 2019, 1,500 employees in Italy, and more than 2,500 worldwide, it represents a global benchmark in the railway world.

With this long history behind me, summarized in a few lines, and in a context such as that of the steel mills, characterized by hard work, sacrifice, sweat and few frills, something even more convincing was needed to give solidity and new life to the entire Lean Lifestyle initiative.

After the critical analysis phase conducted with a small group of executives, it emerged that the biggest obstacle to the evolution of the project was the same that was holding back the momentum toward the excellence of the entire company: the lack of a single corporate improvement plan, which aligned the entire company toward the same goals. Projects arose from the sensitivity of people and functions. Everyone gave their own priorities, and although there were many "good practices" of local improvement management, at the operational level there was little visibility without overall management. This did not favor the synchronization between the various corporate functions and the pursuit of a single, coherent direction for all, with great impact on the continuous involvement of the business leaders.

From this awareness, also fostered by the open discussion with the friends of Streparava S.p.A., during an experience exchange workshop in mid-July 2020, was born the idea of breaking the balance and tradition of the past to organize the Lucchini RS strategic plan in a few months in a complete Hoshin Kanri approach.

"We are here not to become richer, but to create more wealth for us, for those who work with us, for the territory, for the environment and for society," were the words of **Giuseppe Lucchini**, son of founder Luigi, and current president of the company, during the meeting that effectively opened the dances of the entire Hoshin Kanri process on September 3, 2020. On that date, the work plan was presented, which in just four months would have represented a radical change in the company's strategic management model. For the first time, the implementation of the multi-year strategic plan was discussed, involving, from the beginning, a wide audience of contributors who were a bit curious about what was happening.

"The historical period has changed profoundly," says **Augusto Mensi**, "and the long family tradition must be combined in a modern way. Today it is essential to be able to re-interpret the entire company management in a collaborative and participative manner. It is now unthinkable now, to authoritatively impose the will of anyone in the organization. In the last year of working with Lenovys, thanks to the introduction of Hoshin Kanri in the company, there was a great acceleration, triggered precisely by that crisis at the beginning of 2020. In a few months we have achieved alignment towards the company's results by everyone. The fact of not setting goals from above, but to involve and align people in a progressive cascade mechanism, has released energies that I did not imagine. When people understand well why their actions contribute to the realization of an overall design, there is no longer a need to impose. Trust is created and participation becomes spontaneous. You no longer must push for improvement, it is the people who pull and make proposals to help realize the overall design, now clear and visible to all. In this climate, nobody feels offended any more by the confrontations that arise, even intense ones, to achieve a common goal. The role of management also takes on a new and more fulfilling meaning."

In the long work sessions between September and October 2020, all long-term and twelve-month corporate targets were set, divided into three areas: corporate and for each of the two divisions, railway and cast/forged.

After setting targets, all company projects deemed necessary to achieve the objectives, supplementing and deleting where necessary. Between November 2020 and January 2021, all project A3s were formalized, using the format and process seen in Chapter 13. This phase is the one that definitively "hooked" people into the only great improvement plan realized.

"The creation of projects," Augusto Mensi continues, "within the Hoshin Kanri process, managed in a transparent manner and with working groups composed of people from different departments, has reduced the silos effect that previously caused us many problems. Departmental goals and fratricidal struggles have been greatly reduced, and the company's objectives and the projects necessary for the achievement of the company's aim have been set. People open possibilities to review topics that were previously considered taboo. With greater assertiveness. It is easier to question a procedure or process no longer considered functional to the new objectives. When people understand why, they are much more open to dialogue. This is what is happening to us in Lucchini RS. For example, we have recently set up a protected time slot, from 12 to 14 every day, free from meetings and collaborative commitments. Everyone in those two hours can—and should—devote themselves to their Gold Activity, planning, personal growth, strategy. A powerful corporate message, which only a year ago would have been inconceivable to our way of thinking and acting. Before, Lean Management was seen as a tool only for improving efficiency in production departments, now it is becoming more and more a mindset, a cultural approach, a lifestyle."

To facilitate the overall management of the entire process, we have created a digital governance system for the entire Hoshin Kanri strategic plan. From a single dashboard, it is possible to check the real-time progress of each project, access all project A3s, and have visibility of all information and online guides to the methods and tools used. For the executive management of the entire strategic plan, the organization of permanent cross-functional management groups was designed at three levels of organizational depth, from the level where the managing director also participates to the more operational level where only the project operational groups participate. This organization, with the calendar of meetings set for the whole year 2021, aims to constantly monitor both current business criticalities and those of strategic projects. Augusto continues:

I was amazed. We are not Microsoft, we are a Group with steel roots and therefore a reality with its feet firmly on the ground. But now we do things in a way we would never have dared to do before. We start without waiting to be perfect or to have 100% of the information and confirmations in support. Agility for us means being willing to make some mistakes and adapt as we go along, and even though our industry is characterized by heavy investments and choices that have long-term impacts, this direction is required because of the constant change of scenery that we must quickly get used to. If you don't streamline, the market, the competitors, the environment will eat you. From this point of view, the Covid-19 situation has bought us 10 years in the direction of smart working and flexibility of office work. Although most of our people work directly in the production departments, for all the people in the offices, around 250 employees, it is now natural to work on-site or remotely on collaboration platforms, with inter-functional teams, overcoming the concept of functional silos. This way of working has gone hand in hand with the new Hoshin Kanri unified management, in which inter-functional priorities and objectives are defined in advance and clear for everyone.

Methodological support is currently in place for all team leaders of the various strategic projects. The next steps are in the direction of the continuous growth of people, toward the creation of a leadership model increasingly oriented toward the active support of employees to increase their performance and well-being. In a few days we will start the process of reviewing corporate values and related key behaviors, seen as shared rules of the game in corporate interaction. They will be declined and linked to the new performance evaluation system, which will no longer look only at the results, but also at how one works to achieve them.

I look back, I think about those first months of 2020, about that crisis that was about to destroy the whole project at Lucchini RS and I am even more convinced that every company has its own Lean Lifestyle journey to make and that it cannot be copied and pasted, ever. The more we are willing to learn from what happens to us, the more we will grow. Thanks to all the Lucchini RS friends who accepted the challenge and wanted to question themselves, learning together with us.

"Agility, simplification and well-being," Augusto Mensi concludes, "must evolve in unison. People First will be our next big step. We have always been ready to fight, sacrifice, sweat and face the storm. But feeling fulfilled for what you do is the greatest professional success I wish for anyone."

Each Company Has Its Own Journey to Undertake

As you have been able to see from the testimonies and the different stories included in the book, there is in practice no single way of doing Lean Lifestyle projects in a company. There are several possible projects, which need to be "assembled" to measure to allow each company to set out on its own evolutionary journey and achieve results in relation to its starting point and context, in line with the main objective of the Lean Lifestyle: more results and more well-being together.

Some companies have chosen to pursue comprehensive Lean Transformation programs to increase turnover and profits alongside the well-being of their people. Others have chosen to focus on building their business *system* as a unique business communication platform that integrates the core elements of knowledge management with methods and techniques to gain more organizational alignment and more productivity. Still others wanted to complete the organizational redesign and review of their incentive and evaluation system, based on the Lean Lifestyle principles. There are also those who have chosen solely the path of individual and organizational efficiency to reduce stress and extra hours worked.

We have learnt a lot over the years: me, my team, and the thousands of people with whom we have traveled important parts of their personal and corporate Lean Lifestyle journey. There are some recurring elements that have brought more results than others.

The Steps to Embark on Your Journey

If you want to organize your own corporate Lean Lifestyle journey, let me suggest the following steps:

1. sharing the reason for action and high-impact training;
2. envisioning and strategic planning Hoshin Kanri;
3. organization in "inverted pyramid" form;
4. ATRED projects and individual mentoring;
5. corporate projects and mentoring.

Sharing the Reason for Action and High-Impact Training

I remember it as if it were yesterday, my birthday in 2016. I had established a habit that had now lasted for almost two decades: never working on that day.

But on June 27, 2016, I decided to break the habit. I was not doing it out of a sense of duty and because a business offer to an important client was at stake, but because I felt there was something deeper waiting for me. The managing director of the potential client company, after sharing the reasons and aims of the proposed path, had asked me to present the entire Lean Lifestyle corporate initiative to his entire front line that day. If they had shared the reasons and aims of that initiative, we would have left. Otherwise, nothing. Friends as before. I would not have talked about the cost reduction of a product or the efficiency of a production line. Nor about introducing new technology to the company. I would have talked exclusively about how to work better to achieve more performance and well-being. Topics that at first glance can always seem less tangible and more aleatory than others. In the company's main hall, which was made available that day for the presentation, I realized two things:

- When decisions are truly shared, the energy and commitment of people exceeds what you can imagine.
- If you understand what the real problems of the people you work with are, you have already identified a big part of the solution. The rest will be found by them.

At that meeting we tried to understand what the problems were of the people around that table and to share some choices in the direction of their solution. The decision was taken to start with a long management training course that would engage the company front line for eight full-day appointments spread over three months, to be followed by individual and company projects. That company is called Streparava and around that table were professionals who were putting themselves on the line first as people and then as managers: Paolo Streparava, Raffaella Bianchi, Fabio Faustini, Enrico Deltratti, Renato Cotti Piccinelli, Roberto Deltratti, Andrea Ferrari, Stefano Guerra, Davide Ferrario, Roberto Zerbini.

Seeing the passion, energy, and motivation for the journey we were going to take together—in addition to the toast we made together—shine in their eyes, repaid me for "giving up" my long-standing habit of not working on my birthday!

To them I am deeply grateful for the trust placed in them and for the fact that from that date to today we have done so much together.

Why am I talking about high-impact training? Because it is a type of experiential training that aims to prepare the participant for action, called to explore problems and solutions that can have a profound impact on the way he works. The goal is not only to spread knowledge of the principles and operational tools of Lean Lifestyle, but above all to create a strong awareness of the problems and opportunities.

How to choose the type of Lean Lifestyle training most consistent with your needs? In the Streparava example, the company focused on the extensive skills development, and after the comprehensive training done directly in the company for the first line, this was followed by the participation of dozens of other people in the Executive Master Lean Lifestyle Leadership, and the delivery of specific training editions for different company levels. In addition to the corporate or inter-company "live" sessions, more than 200 people had permanent access to our digital Lean Lifestyle Academy platform.

Beyond the specific choices that can be made, it is important to start with one consideration: without competence development, there is no personal and corporate growth. And this applies to Lean Lifestyle competences and all the others that you will need to cope with the changes of the future.

The value of high-impact training is twofold: on the one hand, it aligns the language and basic knowledge in the company, within the scope of the topics covered, and on the other hand, it prepares people for the next step, that of individual and company projects.

> The illiterate of the 21st century will not be those who cannot read and write, but those who cannot learn, unlearn and relearn.
>
> **Alvin Toffler**

> The Lean Lifestyle training has increased in me the awareness that we need to step out of our comfort zones to overcome resistance to change. It is true that a ship in the harbor is safe, but that is not what ships are built for.
>
> **Cesare Ceraso,** Former General Manager of Stanley Black & Decker

Envisioning and Strategic Planning Hoshin Kanri

If there is one company-wide project that has been shown to have more impact than all the others, it is the one that includes the implementation of the envisioning process, followed by the entire Hoshin Kanri corporate strategic planning process, seen in Chapter 13. Why?

The good Lucius Anneus Seneca, back in 65 A.D., almost at the end of his life, stated with proverbial wisdom in one of his Letters to Lucilius that "there is no favorable wind for the sailor who does not know in which port to land." As seen in depth through the pages and testimonies of Chapter 13, it is essential to rediscover and share the sense and meaning of the company and the people who live it every day, to establish the direction that managers and the company want to take, to translate it into the business results that one wants to achieve, and to build a robust system to align all the people and manage the initiatives that contribute to achieving these results in an agile manner. The energy and cohesion derived from this part of the journey are worth all your time invested.

In this way, the specific projects of the Lean Lifestyle journey will not be sporadic or tactical initiatives but can become an integral part of the strategic projects Hoshin Kanri that the company decides to launch. They will therefore be subjected to monthly review, analysis of the results and deviations, like other company work sites.

If a company already has, partially or completely, its own vision-mission-values and strategic planning and management system in place, I suggest that it still devote some attention to the search for anything that could be improved in this area, because, in my opinion, 1% of doubts and misalignments at the top of the company management always translates into 10% of doubts and misalignments at the second level of management, 100% at the third level, and so on. For example, I have often found myself in front of companies that claimed to have a good strategic plan, only to realize, on closer observation, not to have good alignment in the rest of the company and a rigorous and agile execution governance system. Or that they already had their own set of corporate values, only to realize later that they lacked a link to performance evaluation systems.

Organization in "Inverted Pyramid" Form

Beyond how the company is organized, one of the key principles of the Lean Lifestyle is the strong sense of co-responsibility that is transmitted to all levels of the company hierarchy, destroying the old paradigm that can be summed up in the phrase "tell me what to do and I'll do it." The classic pyramidal vision- which sees at the top few who think and at the bottom many who do, following orders- is replaced by an inverted pyramid, using Beppe Scotti's words, in which the many become the thinkers in the company. The entrepreneur and his front line are transformed from givers of orders into inspirers, coordinators, and guides for the rest of the organization, ready to support their people and remove all the obstacles that the various work groups can find along the way.

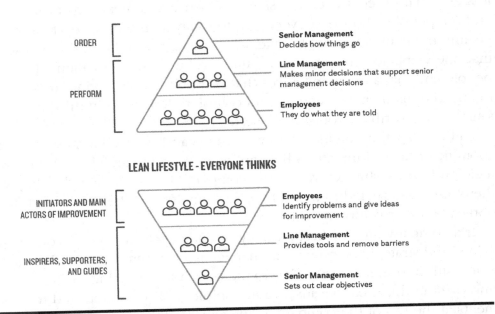

Figure 14.6   Comparison of top-down/bottom-up leadership models.

In terms of design, this concept finds application in the implementation of an organizational system that shifts oversight responsibilities to the intermediate levels of the corporate hierarchy, also delegating the operative decisions as low as possible. This mechanism of organizational delegation works well when behind it there is a robust system of Hoshin Kanri implemented, in which annual objectives and projects have been elaborated and shared by the entire organization, together with the system of government of various projects.

Testimony

If at the beginning the process of change was seen with mistrust, now new proposals are arriving all the time precisely from those who were somewhat more resistant to change. And in my opinion, it has been possible precisely because of this individual and corporate path, which was initially terrifying; and because people now feel directly involved and experience the corporate pyramid as "inverted" compared to before, thanks precisely to the new concept of the centrality of the core staff. Change in the company was frightening at the beginning, and I too was frightened when Luciano asked me if I was sure I wanted to implement such an "invasive" process, which penetrates the organization and disrupts all the established habits.

—**Beppe Scotti**, *CEO Ethos Group*

In Figure 14.7, you can see the example of Streparava where the organizational delegation project led to the creation of a system of committees and functional alignments that ensure permanent supervision from top management to the operational team in production. Committees are permanent cross-functional working groups that meet once or twice a month to make decisions on the two areas of action: current business and strategic projects.

Each company will find its own way and timing to realize the transfer of responsibilities downstream, decide how and how far to go in what is called organizational "empowerment" in some circles.

Based on my experiences over the years, I feel I can say that once we have gone down this road, no one has ever wanted to go back.

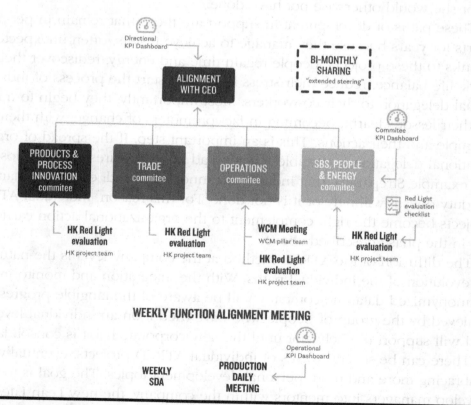

Figure 14.7 Committee system in Streparava.

ATRED Projects and Individual Mentoring

Downstream of each training course chosen, there is almost always a mandatory stage, according to the hundreds of people who have already done it: the launch of an individual project by each of the participants. This project, which lasts approximately 3–5 months, will focus on the application of the ATRED method to reinterpret its organizational role. Inserted in an individual mentoring process, which supports the person in personal development through a mixture of three elements: support with the coaching approach, nudging for change with the nudging approach, and guidance with an experienced mentor who "has already traveled that road before" and who now accompanies you on your own way. In many situations it is important that the coach leaves the person space for reflection and exploration of his solutions through questions. At other times, however, the mentor's ability to propose solutions and tools already validated by experience comes into play, while in other situations it is crucial that the mentor can give that gentle push—nudge—necessary to bring the person to take that extra step, which he or she would otherwise not have done.

These paths of development in support are those that remain in people's hearts for years because they manage to achieve results, often unexpected. Thanks to these projects, people regain time and energy, rediscover their work–life balance, reduce their stress levels and start the process of individual delegation to their co-workers. Most importantly, they begin to transfer their lessons learnt, becoming in fact promoters of change, with their example and their actions. This is an important step. If the spread of organizational delegation is "taxable" creating ad hoc structures such as those of the example Streparava, that individual cannot be forced: either the manager on duty does it willingly or it is not done. For this reason, individual ATRED projects become the right complement to the organizational action carried out in the phase described above.

The diffusion of the ATRED method at the company level is the natural evolution of the individual paths. With the integration and monitoring of anonymized data, the company will be aware of the tangible progress achieved by the group of people who have tried it on an individual level and will support the behavior until the new corporate habit is consolidated.

There can be several waves of individual ATRED projects, eventually embracing more and more personal development topics. The goal is to develop managers into mentors within the company, the new Lean Lifestyle Leaders who can guide and support others in their progress.

Corporate Projects and Mentoring

This is the stage where extensive organizational projects in support of the corporate Lean Lifestyle journey take shape. They can be very different and specific to each situation. In Figure 14.3 you have already seen examples of projects for each type of quadrant. The most frequent and highest impact ones are as follows:

- **Streamlining of key business** processes: these projects are analyzed with a cross-functional value stream approach and never by individual departments. The aim is to simplify and eliminate waste within the analyzed processes to reduce time, increase the value generated and the service level, improve the quality and performance indicators associated with each process and, in the spirit of the Lean Lifestyle, have a positive impact on the well-being of the people involved. An example is the one described in the history of Campari's import/export group in Chapter 8 or those that characterized Orogel's improvement projects in jam production.

- **Dissemination of the key Lean Lifestyle habits:** after selecting the solutions that are considered more in line with the needs and context, we design their operational dissemination to the entire organization. We have seen the example of the organizational habit of ATRED because of the consolidation of the individual habit of its use. Many companies decide to decline slotting techniques and work by cadences through the implementation of a meeting system. Still others decide to set up specific rules and tools for working in sacred time mode—high concentration—by introducing protected time slots, providing soundproofing headphones or allocating special spaces for this purpose.

- **Performance evaluation system:** it is difficult to think about changing people's behavior and habits if we continue to evaluate (or not evaluate) them as we have always done. The evaluation system, already examined in several parts of the book, must be able to capture both performance aspects, but above all those linked to the way people work and consistency with the company's value system. The typical formal annual or six-monthly evaluation, besides needing to be revised, must be supplemented with a more frequent evaluation that meets the need to give close and concrete feedback to people, so that they are guided in their continuous growth.

- **Lean Mailing and Lean Meeting:** we have already explored these types of projects in Chapters 8 and 9. It is a priority in any company to rationalize and reduce the quantity of the two most time-consuming activities.

- **Standardization and construction of the corporate business system:** we covered both these topics in Chapter 7. This is a type of project that, sooner or later, any company should do to ensure organizational alignment, communication, explication of corporate knowledge, operational guidance to everyone active in the organization.

- **Projects to increase the level of energy and work–life balance:** these are projects that act on the different levels of human energy, physical, mental, and emotional, with the specific aim of educating people to "take care of themselves" in the full balance of working and personal life.

- **Projects of human development and emotional energy growth:** consist in the accompaniment for the evolution of the leadership model, for example from authoritarian head-to-head coach, able to develop himself and others, increasing the level of emotional energy in the company. The criteria of Habit Management are applied in this type of project, to accompany people in acquiring new habits functional to the growth of emotional intelligence.

- **Smart layout:** as in the Timenet example described in Chapter 7, this type of project aims to remove environmental obstacles to agile work, to encourage highly concentrated work, to allow moments of communication in small groups, to allow the flexibility required by agile and remote working. Through this type of project, efforts are always made to increase people's energy.

- **Setting up remote agile working models:** accompanying the company in the full implementation of what was seen in Chapters 11 and 12. Depending on the specific starting point, a program is set up to operationalize all the conditions for an agile work, which is not only "remote" but becomes a more evolved form of agile working.

- **High-impact innovation pathways:** after the initial training on high-impact innovation topics, we proceed with the accompaniment in the implementation of pilot projects to generate, validate, develop, and bring to market new high-impact innovative solutions, for example new products or services, with their business models in support.

When I chose the Lean Lifestyle approach within engineering at ELT Group in 2017, I saw a double opportunity: to grow people in order to face the challenges in front of us and at the same time be more attractive as a company. They liked the approach and we have had huge success. I see much more energy and enthusiasm in people: we are back to our roots, when our founder used to say, "without enthusiasm there is nothing."

Many habits in the company have changed and the Lean Lifestyle has given us a push in this direction. Just think, for example, of the impact that has had for engineering and production has had on the lead time of 36 months to 12–18 months. We are still in the middle of the journey and right now the challenge for us is to get the other functions on board.

The direction, however, is clear and it is no coincidence that in the Hoshin Kanri we have just developed in Engineering, one of the strategic goals we have chosen is to become a Lean Lifestyle Company.

Simone Astiaso, *Vice President Engineering &*
Operations at ELT Group

Slow Down and Reflect

There Are Times When You Must Slow Down and Reflect, to Choose to Act Differently

We are at the end of the book, and I think this may be one of those.

If you have come this far, you are probably also pondering how to act and how to overcome small or big challenges to improve your professional living conditions and those of the people you work with.

Through the way we work and the way we do business we all face daily changes and challenges.

Companies, in this changing historical phase, must and can play a strategic role in the human, social and economic development of our country. And they can do much to implement technological and management systems that help people achieve high forms of productivity, human and corporate excellence, and individual and collective well-being.

I believe it is possible to achieve short-term results and at the same time cultivate great long-term business dreams, creating prosperity and value for working people, society, the environment, the market, and shareholders. Without compromise.

I believe it is possible to achieve high levels of business efficiency and productivity while simplifying and streamlining people's working days. Without compromise.

I believe it is possible to move the organization with efficiency and agility and at the same time have people who feel fulfilled, energized, and involved in the company strategy. Without compromise.

I believe it is possible to pursue corporate excellence and at the same time cultivate excellence as a way of life outside the company, without having to choose whether to prioritize work or personal life. Without compromise.

Together with my team and based on the experiences of thousands of people who have shared these values over the years, we have conceptualized and disseminated a company model: the Lean Lifestyle Company. A company that aims to create maximum value and express its full human, technological and organizational potential. A company that does not passively look at what is happening around it but is committed to governing the current complexity and addressing the changes taking place to achieve great performance; at the same time, it takes care of its own internal well-being, aware that this will also bring well-being outside the company itself.

In conclusion and as a summary of this book, I want to share the Lean Lifestyle Manifesto where the key values and behaviors that characterize a Lean Lifestyle Company are collected and summarized. A real collection of best practices and operating principles for those who want to evolve the way they work and do business in the current world.

The Manifesto aims to unite entrepreneurs and managers who share the awareness that every choice they make always has an impact. Because through the evolution of the way we do business, we can leave a tangible mark that starts from the company boundaries, touches ourselves and the people we work with, and even positively influences society and the environment in which we live.

To take ourselves, companies, and society to new and unexpected heights.

The Lean Lifestyle Manifesto

1. **Vision and execution**
   We want to create a company that encourages people to think, dream, and achieve something big. A company that always involves and aligns

all employees. A company that grows and develops based on what it has built in the past, but at the same time cultivates a strong orientation toward the future with long-term strategies and new, ambitious goals. A company that pays attention to the impact of its actions on people, society, and the environment. For the realization of corporate strategies, we want to create systems of involvement extending to the entire organization so that everyone can contribute. *Dreaming big and acting small every day* is what has enabled the great achievements of humanity. That is why we will measure and value short-term progress toward long-term goals: in this way, every day we will take a small step toward the desired direction.

2. **High-impact innovation**

We never want to stand still and passively watch what is happening around us. We will *innovate to change what exists for the better: processes, products, technologies, and business models.* And we will do it with high impact, not chasing fashions, trends, and last-minute inventions, but trying to achieve tangible impacts for the market and the company itself with the lowest possible cost and in the shortest possible time, in an agile mode. In an era of great and rapid change, a Lean Lifestyle Company continually prepares its people to embrace with curiosity all those evolutions that bring greater value to customers, greater flexibility and agility to companies, and better quality of life to people. It is a company that always has a reservoir of structured ideas ready to become solutions. It will never have an innovation project of a single person or a single department but always the business project of a multidisciplinary team that from the outset methodically tackles the various stages of development, from the idea to the solution launched on the market.

3. **Continuous experimentation for a *data-driven company***

In a Lean Lifestyle Company, we do not judge a new solution based on beliefs, preconceptions, or what one has always done in the past. One accepts a priori that one may be wrong and *experiments to decide what is best for oneself, one's customers, and the company,* choosing on the basis of facts and tangible data, in essence becoming a "data-driven company." The logic of experimentation is sought in every field: from changing or improving a process to the launch of a new product or service, and from adopting a new habit to the choice of a new investment.

Understanding how to design and customize each experiment is the real key to success. The more people learn cyclically to design experiments, implement, and learn, the more they can embrace personal and corporate progress.

4. **Focus on value**

   Those who work in a Lean Lifestyle Company know what they want, and, above all, what matters: they do not like to waste their energy. That is why we want to develop a company model that always helps its people to express the best of themselves and to bring out their true potential. A company in which the few priority and important activities are periodically selected: those that have the maximum impact in the realization of the company's objectives and that manage to generate the highest level of passion and enthusiasm in people and the organization.

5. **Agility and simplification**

   A Lean Lifestyle Company always seeks the most efficient, faster, less complex, and less bureaucratic way of working, in compliance with the relevant regulations. In this way, available energy, time, and resources are conserved, guaranteeing speed of execution and response to customers. The company's corporate infrastructure, layouts, and physical locations are geared to ensure efficiency and streamlined operations. It is a well-established practice to go hunting for waste in all business processes, remembering well that initial rush and lack of synchronization often leads to rework, subsequent loss of time, and longer overall timescales in any field.

6. **Orientation to results**

   In a Lean Lifestyle Company, there is a strong awareness that in the end *all that remains of the work done are the results and impacts we have generated on people, on customers, on profits, on the environment, on society*, and not the many activities that are done daily. This is why behavior geared toward always improving current results is encouraged and time-wasting, sterile polemics, indecisions, and activities not aimed at improvement are discouraged. In this way, the time required to carry out activities and projects in the company will be reduced even further, maximizing the generation of value for internal people and external clients.

7. **Digitizing to create value**

Digitizing processes that are not agile to start with is a danger that many organizations run, because automating waste will always amplify the negative consequences for people and companies. *A Lean Lifestyle Company exploits all the levers of digitization to generate more value and reduce waste both to external customers and in internal processes.* Digitization is adopted to help people make timely decisions supported by facts and figures available in real time, avoiding redundant operational steps and unnecessary time loss. This approach fosters the right digital mindset: think value first, then the processes and organization required to deliver it. In this way, digitization will be the basis for getting the entire business system off the ground through truly lean and value-oriented processes.

8. **Smooth and value-oriented organization**

Lean Lifestyle key values and behaviors can only be acted on and lived on easily if systems and an organization have been created in the company that enable change and support new habits over time. The end customer does not care how we are organized, but only how much value we receive from our company and how quickly we respond to their needs. The culture of functional silos often generates unnecessary conflict and does not guarantee maximum value in the shortest times. *A Lean Lifestyle Company adopts value-oriented organizational models, with multidisciplinary, autonomous, and fast teams*, with people having full responsibility for the process or inter-functional project entrusted to them.

9. **Development of full human potential**

In a Lean Lifestyle Company, we value the people we work with and ask ourselves every day how we can best develop the potential of each employee. This is why we make extensive use of delegation and delegation as pivotal tools for professional development: to make managers grow, to free up time and energy to devote to new activities, and to make other employees grow by entrusting them with new responsibilities. We build *an army of problem solvers in the company, autonomous and flanked by their managers for ongoing support and development*. Because the necessary condition for the growth of companies is precisely the growth of them. In this way everyone learns to have more and more confidence in themselves and in others. What the individual can *do is never comparable with what a cohesive group of people can achieve.*

10. **Physical and mental well-being**

We believe that professional performance and the overall well-being of people are two sides of the same coin, which only perform best when they act synergistically. One helps to improve the other. This is why we encourage and disseminate culture and structured actions to increase awareness, sensitivity, and training of people in this area, helping them to take care of themselves, increase their energy, and regain the power of concentration. In the continuous improvement of a Lean Lifestyle Company's business processes, therefore, we will always try to understand what can make performance grow and at the same time make people more active and vital, while minimizing anything that can drain energy. *The goal is to achieve higher quality, sustainable human performance in the company combined with increasing forms of physical and mental well-being.*

11. **Emotional energy and true relationships**

In a Lean Lifestyle Company there is a deep awareness of the importance of people's emotional energy. In fact, despite all the technological progress, human beings are activated and guided in their behavior by what they feel: their emotions. For this reason, *everything that generates positive emotional energy is encouraged in the company. Small and great successes are celebrated, thanks are given for the work done without ever taking it for granted*, improvements are appreciated according to the individual's starting level, and the value of employees is recognized. Active and attentive listening of people is encouraged, and feedback is provided in a constructive and incisive manner. In this way, true human relationships are built in the company, the stable links of a cohesive, strong, and flexible chain.

12. **Pursuit of excellence as lifestyle**

As entrepreneurs and managers of a Lean Lifestyle Company, *we promote the pursuit of integral excellence in all areas of life, inside and outside the company.* This will inevitably lead to harmonizing the overall development of the whole person, helping to make them an example of excellence. We will help people, even outside the company boundaries, to achieve the highest balance in life, encouraging them to continuously improve and learn new skills to continue to evolve, inside and outside the company. In this way, we will develop adaptability and resilience in a rapidly changing world and seek an overall harmony between personal and working life, to achieve excellent and sustainable human performance over time.

I hope this book can be a guide and inspiration for you and for anyone who wants to embark on the journey toward the realization of his Lean Lifestyle Company.

If you have recognized yourself in the principles of its Manifesto, if you have already implemented one or more of the elements described and wish to share your experiences, but above all if you accept the challenge to contribute to make companies places where great results and great well-being can be achieved at the same time, places where people and companies can grow in unison, then I invite you to join the Lean Lifestyle Community on www.lucianoattolico.com website to stay connected with me and all the people who are making their own Lean Lifestyle journey.

Thank you.

Luciano Attolico

Index

Note: Page numbers in *italic* indicate a figure on the corresponding page.